MULTILINGUAL GLOSSARY
OF
AUTOMATIC CONTROL TECHNOLOGY

English · French · German · Russian
Italian · Spanish · Japanese · Chinese

SECOND EDITION
1995

INTERNATIONAL FEDERATION OF AUTOMATIC CONTROL

IFAC, the International Federation of Automatic Control, comprises forty-eight national member organizations which collaborate through a structure of technical committees – forty-six at present – to develop and promote all aspects of the subject of control. IFAC holds around forty technical meetings each year, in two principal formats, symposia and workshops, plus a triennial congress. IFAC journals and conference volumes are published by Elsevier Science Ltd, The Boulevard, Langford Lane, Kidlington, Oxford OX5 1GB, England.

AUTOMATICA

Automatica has long been established as a leading journal of control. *Automatica* is published monthly and is edited by an international group of editors and associate editors. The journal publishes research papers and surveys on all aspects of theoretical and experimental research and its application to all types of control system.

CONTROL ENGINEERING PRACTICE

Control Engineering Practice is a monthly journal for engineers whose research deals with real life problems of control, largely in industry. The journal contains papers with a strong applications flavour, centred in engineering but extending into applied fields to which control is relevant. Papers contain the results of design and data from real systems. Submissions from industry are encouraged.

IFAC PROCEEDINGS AND POSTPRINT VOLUMES

IFAC Postprint Volumes contain the papers presented at IFAC symposia and at selected workshops. The papers from the triennial congress of IFAC are published as *Proceedings*.

For information about IFAC, including society and individual membership, contact:

IFAC Secretariat
Schlossplatz 12   Telephone: +43-2236/71 4 47
A-2361 Laxenburg   Telefax: +43-2236/72859
Austria   e-mail: ifac@serv.univie.ac.at

For information about publications, contact:

Elsevier Science Ltd
The Boulevard   Tel: (+44) (0) 865 843000
Langford Lane   Fax: (+44) (0) 865 843010
Kidlington
Oxford OX5 1GB
England

MULTILINGUAL GLOSSARY
OF
AUTOMATIC CONTROL TECHNOLOGY

English · French · German · Russian
Italian · Spanish · Japanese · Chinese

Editors
H. A. Prime
A. Work

Compilers

J. Beneš	G. A. Ferratè
P. Gilard	J. Hoffmann
M. Masubuchi	R. Oetker
H. A. Prime	T. Raimondi
Z. Wang	A. Work

Z.-Q. Wu

Assisted by
H. Akashi, P. Albertos, A. Bastow, R. Chaussard, H.-F. Chen,
J. F. Coales, G. Guardabassi, A. Kuznetzov, S. Li, Y.- Z. Lu and M. Thoma

IFAC

Published in collaboration with the

INTERNATIONAL FEDERATION OF AUTOMATIC CONTROL

under the guidance of the Technical Committee on Terminology and Standards

PERGAMON
An imprint of Elsevier Science

U.K.	Elsevier Science, Ltd, The Boulevard, Langford Lane, Kidlington, Oxford, OX5 1GB U.K.
U.S.A.	Elsevier Science, Inc., 660 White Plains Road, Tarrytown, New York, 10591-5153, U.S.A.
JAPAN	Elsevier Science Japan, Tsunashima Building Annex, 3-20-12 Yushima, Bunkyo-ku, Tokyo 113, Japan

First edition 1981
Second edition 1995

Library of Congress Cataloging in Publication Data
A catalogue record for this book is available from the Library of Congress

British Library Cataloging in Publication Data
A catologue record for this book is available from the British Library

ISBN 0 08 037192 2

Transferred to digital printing 2005
Printed and bound in Great Britain by Antony Rowe Ltd, Eastbourne

CONTENTS

Foreword

Twenty years have passed since the first IFAC Multilingual Dictionary of Automatic Control Terminology was published in 1967. It was during the 2nd IFAC Congress held in Basle in 1963 when the first proposal was made to start preparations for editing this six-language dictionary, and the responsibility was taken by a newly elected Committee. The late Professor D T Broadbent acted as the Chairman of the Committee and he was also the General Editor of the first IFAC Dictionary.

The main emphasis was placed on the basic concepts of automatic control and the related terminology. A limited number of closely related terms for components and also a few terms from the field of computer technology were included.

The first Dictionary consisted of a Vocabulary in which words and expressions were arranged according to technical topics, and six alphabetical indexes which permitted the location of a given term in the Vocabulary.

The discussions of the necessity of revising the first IFAC Dictionary started in the IFAC Technical Committee on Terminology and Standards (TC on TERMST) in 1972. The proposition by the former Chairman Dr H L Mason to add Japanese terms was earnestly supported and adopted by the next Chairman Professor J Benes. In 1980 it was decided to add Japanese terms to the original edition of 1967 and publish the Dictionary as a seven-language glossary by PERGAMON PRESS in time for the IFAC Kyoto Congress in 1981. This plan was completed in due time thanks to the very effective work of Professor M Masubuchi, the Editor for Japanese (using both Kanji and Kana characters).

Meanwhile, Professor Benes edited PRODROMUS I and II, the lists of new terms on automatic control, collected by the members of the TC on TERMST and proposed for inclusion into the new IFAC Dictionary.

Since the Kyoto Congress in 1981 the present Chairman of the TC on TERMST has been responsible for collecting and selecting new terms in English for the updated IFAC Dictionary, which were listed in the working materials PRODROMUS III to VI.

What we have reached is the result of mostly collective work of the active members of the TC on TERMST. Sometimes it was difficult to decide whether to include or omit a term. The present choice may be very subjective. However, a work like this can never be completed. The basic ideas while selecting for inclusion new terms on automatic control were the following: the terms should not be too general or too specific (used only in certain applications); there must not be too long; and there should be only a minimum of computer technology terms, otherwise the number of terms in the Dictionary would increase enormously. Besides, the latter field of terminology is covered by IFIP among other international bodies.

During the IFAC Budapest Congress in 1984, several important (from our point of view) decisions were made by the TC on TERMST. Firstly, the new IFAC Multilingual Dictionary would have eight languages (including Chinese).

Secondly, all the terms would be arranged alphabetically. Thirdly, where advisable, the terms would have short explanations or definitions in English.

After the Budapest Congress, a subcommittee checked the IFAC Dictionary (1981 edition), and as a result of their work almost 100 terms were considered unnecessary in the new Dictionary. During 1984, Professor H A Prime compared the old IFAC Dictionary (878 terms) and PRODROMUS VI (540 terms) with the INSPEC THESAURUS (1983 edition), and found many additional terms. After

discussions with the TC members, a list of 142 terms was made up as the ADDITION to PRODROMUS VI.

By the end of 1985 the final set of English terms for the new IFAC Multilingual Dictionary was ready. The translation of the terms into other languages was the responsibility of the Language Editors (G A Ferraté for Spanish, J Hoffmann and P Gilard for French, M Masubuchi for Japanese, R Oetker for German, T Raimondi for Italian, A Work for Russian, Wu Zeng-Qian for Chinese) and this job was completed in October, 1986.

The intention to add some short explanations or comments to several terms was abandoned because of the suggestion made by the IFAC Publications Managing Board keeping in mind their future plans of publishing a multivolume Pergamon Control Encyclopedia.

Now, after many years of preparatory work the new IFAC Multilingual Dictionary of Automatic Control is published. I would first of all like to thank Professor T Vamos and Professor M Thoma, Past Presidents of IFAC, for their permanent interest in this work. I would like to thank Professor J F Coales, Past Chairman of the IFAC Publications Managing Board, for his share in giving the final shape to the Dictionary.

I would also like to thank all the active members of my Technical Committee for preparing the final set of terms in English and all the Language Editors for translating the terms into their native languages.

I would particularly like to thank my secretary, Mrs O Uibo who carefully typed almost all the working materials. I am also grateful to my colleagues from the Institute of Cybernetics and the Tallinn Technical University who made use of the text processing systems and helped me in translating the terms into Russian. The aid of my former instructor, Professor H Sillamaa, was indispensable.

I want to express my sincere gratitude to everybody whose assistance has made the publication of this Dictionary possible. For me it has been a pleasure to participate in this work.

As I mentioned previously, a dictionary of a rapidly developing field can never be complete. The compilation of one is rather a never-ending process intentionally stopped for a while in order to record the terminology at a particular moment and, without delay, make it accessible to many people with different native languages, thus contributing to a better understanding in the world (at least) of automatic control. At the same time it is also meant to influence the languages where the respective terminology is just beginning to develop.

I trust the next IFAC Dictionary will be published before another 20 years pass and that it will also include those terms already created or just about to appear which we were not able to include in the present Dictionary.

All critical remarks and suggestions will be greatly appreciated and may be addressed to the IFAC Secretariat (Schlossplatz 12, A-2361 Laxenburg, Austria).

Ants Work
Chairman, IFAC TC on TERMST

Postscript to the Foreword

Professor Ants Work completed his term of office as Chairman of the TERMST Committee of IFAC on the conclusion of the Munich Congress in 1987. His outstanding work as Chairman of the TERMST Committee will be recognised in the publication of this second edition of the Multilingual Dictionary. Professor Work was succeeded in 1987 by Professor Solheim who would undoubtedly have continued to maintain the progress of the editorial work had it not been for his most untimely death so shortly after the 1987 Congress. As Vice-Chairman of TERMST, I was then invited to take over the Chairmanship of TERMST and Professor Otteblad became Vice-Chairman. We both owe much to the standards set by Professor Work and it has been our primary objective to maintain those standards.

The development of the control and systems disciplines has never been static and this is manifest in this Second Edition by the fact that it includes more than 1500 terms. The increasing interaction between the control and computing sciences is also evident in the new edition, but only those terms that clearly have a dual context have been included.

The format of the Second Edition is also significantly different to that of the original edition in that the numerical listing is entirely alphabetical for the English terms and no attempt has been made to group terms within a particular context, as in the previous edition. This change has been made in order to comply with the methods of presentation adopted in other multilingual glossaries in other technical disciplines. The translations into the other seven languages are then listed in columns adjacent to the English listing. An alphabetical index of the terms in each of the other seven languages is included in order to provide the listed number of the equivalent English translation. It is intended that this method of presentation should provide a more effective and quicker means of translation.

In conclusion, I would wish to express my personal thanks to all who have been involved in the project, whether as members of IFAC in providing translation or other editorial material, or as members of Pergamon Press in Oxford, with whom I have had a close association over the past four years.

H A Prime

Vingt années se sont écoulées depuis la publication, en 1967, du premier dictionnaire multilingue IFAC sur la terminologie du contrôle automatique.

C'est au cours du deuxième congrès de l'IFAC, à Bâle en 1963, que l'on fit la première proposition de commencer la préparation d'un dictionnaire en six langues sous la responsabilité d'un comité nouvellement créé.

Le défunt professeur D T Broadbent on fut à la fois le président et l'éditeur du premier dictionnaire IFAC.

Le contenu reposait principalement sur les concepts de base du contrôle automatique et de la terminologie afférente. Un nombre limité de termes relatifs aux composants ainsi que quelques termes relevant du domaine de la technologie des calculateurs y furent inclus.

Le premier dictionnaire consistait en un vocbulaire dans lequel les termes et les expressions étaient classés par objets techniques et de six index alphabétiques permettant de retrouver un terme dans le vocabulaire.

Les discussions sur la nécessité de réviser le premier dictionnaire IFAC commencèrent en 1972 au sein du Comité Technique IFAC pour la Terminologie et la Normalisation (Comité TERMST). Le nouveau Président, le Professeur J Benes soutint et adopta la proposition de son prédécesseur, le Dr. H L Mason, d'inclure les termes en japonais. En 1980, on décida d'ajouter les termes en japonais à l'édition originale de 1967 et de fair publier le dictionnaire par PERGAMON PRESS sous forme d'un glossaire en sept langues en vue du Congrès de Kyoto en 1981.

Ce plan fut réalisé à temps grâce au travail efficace du Professeur M Masubuchi, éditeur pour le japonais (en utilisant les deux types de caractères Kanji et Kana).

Entretemps, le Pr. Benes édita PRODROMUS I et II, deux listes de nouveaux termes en contrôle automatique, rassemblés par les membres du Comité Technique TERMST et proposés pour insertion dans le nouveau dictionnaire IFAC.

Depuis le Congrès de Kyoto, en 1981, le Président en exercice du Comité Technique TERMST fut responsable de la collecte et de la sélection de nouveaux termes en anglais en vue de la mise à jour du dictionnaire IFAC, termes qui furent regroupés dans les documents de travail PRODROMUS III à VI.

Ce que nous avons réalisé est le résultat du travail collectif des membres actifs du Comité. Il fut parfois difficile de décider d'accepter ou d'exclure un terme. Le choix actuel peut paraître subjectif. Toutefois, pareil travail ne peut jamais être terminé.

Les idées de base conditionnant la décision d'inclure de nouveaux termes relatifs à la régulation automatique furent les suivantes: les termes ne doivent être ni trop généraux ni trop spécifiques (c'est-à-dire utilisés seulement dans certaines applications ils ne doivent pas être trop longs; il ne doit y avoir qu'un nombre minimal de termes relatifs à la technologie de l'informatique sinon le nombre de termes du dictionnaire aurait considérablement augmenté. Par ailleurs, ce dernier domaine de la terminologie est couvert par l'IFIP parmi d'autres organisations internationales.

Lors du Congrès de Budapest, en 1984, plusieurs décisions importantes (à notre point de vue) furent prises par le Comité Technique TERMST. Premièrement, le nouveau dictionnaire multilingue de l'IFAC serait en huit langues (y compris le chinois). Deuxièmement, tous les termes seraient classés par ordre alphabétique. Troisièmement, si nécessaire, les termes seraient brièvement expliqués en anglais.

Après le Congrès de Budapest, un sous-comité contrôla l'édition de 1981. Il conclut devoir supprimer environ 1000 termes dans le nouveau dictionnaire. En 1984, le Professeur H A PRIME compara l'ancien dictionnaire IFAC (878 termes) et le PRODROMUS VI (540 termes) avec le THESAURUS INSPEC (édition de 1983) et y trouva bon nombre de termes supplémentaires. Après discussion avec les membres du Comité, une liste de 142 termes fut établis en tant que supplément au PRODROMUS VI.

A la fin de 1985, une liste finale de termes anglais était prête. La traduction des termes dans les autres langues fut confiée au "éditeurs linguistes" (G. A. FERRATT pour l'espagnol, J. HOFFMANN et P GILARD pour le français, M MASUBUCHI pour le japonais, R ORTKER pour l'allemand, T RAIMONDI pour l'italien, A WORK pour le russe et WU ZENG-QIAN pour le chinois. Le travail fut achevé en 1986.

L'intention d'ajouter quelques brèves explications ou commentaires à certains termes fut abondonnée à la suggestion du Comité Directeur des publications de l'IFAC sur la base de ses projets de publier une encyclopédie PERGAMON sur le contrôle en plusieurs volumes.

Maintenant, après bien des années de travail préparatoire le nouveau dictionnaire multilingue IFAC est publié.

Je voudrais, en tout premier lieu, remercier les Professeurs T VAMOS et M THOMA, anciens présidents de l'IFAC de leur intérêt permanent pour le travail. Je voudrais aussi remercier le Professeur J F Coales, ancien président du Comité de Direction des publications de l'IFAC d'sa participation dans la présentation finale du dictionnaire. Mes remerciements s'adressent également à tous les membres actifs de mon Comité Technique pour la préparation de la liste finale des termes en anglais ainsi que les traducteurs dans les différentes langues.

Je voudrais aussi remercier spécialement ma secrétaire, Mme O UIBO qui a soigneusement dactylographié tous les composants du travail. Ma reconnaissance va également à mes collègues de l'Institut de Cybernétique et de l'Université Technique de Tallinn qui utilisèrent les systèmes de traitement de texte et m'aidèrent à traduire les termes en russe. L'aide de mon ancien maître, le Professeur H SILLAMAA était indispensable.

J'exprime enfin toute ma reconnaissance à tous ceux qui rendirent possible la publication de ce dictionnaire. Ce fut pour moi un grand plaisir de participer à ce travail.

Ainsi que je l'ai mentionné précédemment, un dictionnaire relatif à un domaine en rapide expansion ne peut jamais se terminer.

La compilation des termes est plutot un processus sans fin, arrêté intentionellement pendant un certain temps pour enregistrer la terminologie à un moment donné et la rendre accessible sans délai à un grand nombre de gens parlant différentes langues, contribuant ainsi à une meilleure compréhension, au moins dans le monde du contrôle automatique. En même temps, on pense à influencer les langues où une terminolgie propre est en train de se développer.

Je suis persuadé que le prochain dictionnaire IFAC sera publié avant que ne passent vingt autres annés et qu'il contiendra les termes déjà créés ou prêts à l'être que nous ne fumes pas en mesure d'inclure dans cette édition.

Toute critique ou suggestion qui sera très appréciée peut être adressée au Secrétariat de l'IFAC (Schlossplatz 12, A-2361 LAXENBURG)

Ants WORK
Président du Comité TERMST de l'IFAC

Complement a la Préface

Le Professeur Ants WORK a achevé son mandat de Président du Comité TERMST de l'IFAC à la clôture du Congrès de Münich en 1987. Son remarquable travail se reconnaîtra dans la publication du dictionnaire multilingue. Au professeur WORK succéda, en 1987, le Professeur SOLHEIM qui, incontestablement, aurait poursuivi le travail d'édition ne fut-ce son décès inopiné après le congrès de 1987. En ma qualité de Vice-Président du 'TERMST, je fus invité à en assurer la Présidence, le Professeur OTTENBLAD devenant Vice-Président. Tous deux, nous d vons beaucoup aux normes établis par le Professeur WORK et notre premier objectif fut de les conserver.

Le développement des techniques du contrôle et des systèmes ne fut jamais statique et ceci est manifeste dans cette seconde édition par le fait même qu'elle contient plus de 1500 termes.

L'interaction croissante entre les sciences de l'Informatique et de la Régulation Automatique apparaît d'évidence dans la nouvelle édition mais, suels les termes apparaissant dans un double contexte ont été retenus.

Le format de la seconde édition est aussi nettement différent de celui des éditions précédents en ce sens que la liste numérique anglaise est entièrement alphabétique et que l'on n'a pas essayé de grouper les termes dans un contexte particulier comme ce fut le cas dans la première édition.

Cette modification est intervenue en vue de se conformer aux méthodes de présentation adoptées dans d'autres dictionnaires multilingues relevant d'autres disciplines techniques.

La traduction dans les sept autres langues est ensuite présentée en colonnes adjacentes à la liste anglaise. Un index alphabétique des termes dans les sept autres langues est inclus en vue d'indiquer les numéros des termes correspondants de la liste anglaise.

Cette présentation a pour but d'accélérer plus efficacement la traduction.

En conclusion, je souhaite exprimer mes remerciements personnels à tous ceux qui participèrent au projet, qu'ils soient membres de l'IFAC en ayant fourni des traductions ou tout autre matériel d'édition ou membres de PERGAMON PRESS à Oxford avec qui j'ai collaboré au cours des quatre dernières années.

H A Prime

Vorwort

Zwanzig Jahre sind vergangen seitdem das Erste IFAC mehr sprachige Wörterbuch für Begriffe der automatischen Steuerung und Regelung 1967 veröffentlicht wurde. Es war während des zweiten IFAC-Kongresses in Basel, als der erste Vorschlag gemacht wurde, die Vorbereitungen zu beginnen, um die ses sechs-sprachige Wörterbuch herauszugeben und ein neu gewähltes Kommittee übernahm die Verantwortung. Der verstorbene Professor D T Broadbent wirkte als Vorsitzender des Kommittees und war auch der Hauptherausgeber des ersten IFAC Wörterbuchs.

Hauptsächlich behandelt wurden die grundlegenden Konzepte der automatischen Steuerung und Regelung, sowie die diesbezüglichen Begriffe. Einbezogen wurden eine begrenzte Anzahl einschlägiger Begriffe der Gerätetechnik und auch einige Begriffe aus dem Gebiet der Rechnertechnik.

Das erste Wörterbuch besteht aus einem Vokabular, in dem Wörter und Ausdrücke dem technischen Inhalt entsprechend angeordnet sind und sechs sprachlichen Inhalts verzeichnissen, die es gestatten den Ort eines gegebenen Begriffs im Vokabular zu finden.

Aussprachen über die Notwendigkeit einer Überarbeitung des ersten IFAC Wörterbuchs begannen im Jahre 1972 im "IFAC Technical Committee on Terminology and Standards" (TC on TERMST). Der frühere Vorsitzenden Dr. H L Mason, japanische Vorsitzenden Professor J Benes. Im Jahre 1980 wurde beschlossen, japanische Begriffe in die ursprüngliche Ausgabe von 1967 einzufügen und das Wörterbuch als sieben-sprachige Ausgabe durch die PERGAMON PRESS gleichzeitig mit dem IFAC Kyoto Kongreß 1981 bekannt zu geben. Dieser Plan wurde zur rechten Zeit erfüllt, dank der sehr Wirkungsvollen Arbeit von Professor M Masubuchi, dem Herausgeber für Japanisch (Sowohl Kanji als auch Kana-Schriftzeichen benutzend).

Inzwischen gab Professor Benes PRODROMUS I und II, Listen neuer Begriffe der Steuerung und Regelung, heraus, die von den Mitgliedern des TC on TERMST gesammelt und vorgeschlagen wurden zur Aufnahme in das neue IFAC-Wörterbuch. Seit des kyoto kongresses 1981 ist der gegenwärtige Vorsitzende des TC on TERMST verantwortlich gewesen, neue Begriffe in Englisch für das obige Wörterbuch zu sammeln und auszuwählen, sie wurden in den Arbeits unterlagen PRODROMUS III bis VI aufgelistet.

Was wir erreicht haben ist das Ergebnis der vorwiegend sammelnden Arbeit der aktiven Mitglieder des TC on TERMST. Zuweilen war es schwierig zu entscheiden ob ein Begriff aufzunehmen oder auszuscheiden ist. Die gegen wärtige Auswahl scheint ziemlich subjektiv zu sein, jedoch eine Arbeit wie diese ist niemals vollständig. Die Grundgedanken bei der Auswahl neuer Begriffe der automatischen Steuerung und Regelung zur Aufnahme waren die Folgenden: Die Begriffe sollten nicht zu allgemein und nicht zu speziell (nur bei gewissen Anwendungen gebräuchlieb) sein; sie dürfen nicht zu lang sein; und es sollte hier nur ein Minimum von Begriffen der Rechner-Technik geben, sonst hätte sich die Anzahl der Begriffe im Wörterbuch enorm vermehrt. Außerdem wird dieses Gebiet der Terminologie durch die IFIP in anderen internationalen Körperschaften behandelt.

Während des IFAC-Kongresses in Budapest wurden verschiedene, unseres Erachtens wichtige Beschlüsse vom IC on TERMST gefaßt. Erstens: das neue mehrsprachige IFAC-Wörterbach soll acht Sprachen, einschließlich Chinesisch, enthalten.

Zweitens: alle Begritte sollen alphabetisch angeordnet werden. Drittens; wo es ratsam erscheint, sollen Begriffe kurze Erklärung und Definition in English erhalten.

Nach dem Budapester Kongreß prüfte ein Unterkommittee die Begriffssammlung von 1981 für das IFAC Wörterbuch und als Arbeitsergebnis wurden beinahe 100 Begriffe für die Neuausgabe als unnötig betrachtet. Während 1984 verglich Professor H A Prime das atte Wörterbuch der IFAC

(878 Begriffe) und PRODROMUS VI (540 Begriffe) mit dem INSPEC THESAURUS (Ausgabe 1983) und fond viele zusätzliche Begriffe. Nach Rücksprache mit den Mitgliedern des TC on TERMST, wurde eine Liste von 142 Begriffen als Zusatz (ADDITION) zu PRODROMUS VI festgelegt.

Ende 1985 wer die endgültige Sammlung Englischer Begriffe für das neue mehrsprachige IFAC Wörterbach fertig. Die Übersetzung der Begriffe in andere Sprachen war die Verantwortlich keit der sprachlichen Herausgeber: G A Ferrate Spanisch, J Hoffmann und P Gilard Französisch, M Masabuchi Japanisch, R Oetker Deutsch, T Raimondi Italienisch, A Work Russisch, Wu Zeng-Qian Chinesisch, und diese Arbeit wurde im Oktober 1986 vollendet.

Die Absicht einige kurze Erklärungen oder Anmerkungen verschiedenen Begriffen hinzuzufügen wurde aufgegeben, da seitens des IFAC Publications Managing Board angeregt war, ihre zukünftigen Pläne der Veröffentlichung einer mehrbändigen Pergamon Steuerungs und Regelungs Enzyklopodie beizabehalten.

Nun, nach mehreren Jahren der Vorbereitungsarbeit ist das neue mehrsprachige IFAC Wörterbuch der automatischen Steuerung und Regelung zu veröffentlichen. Zu allererst möchte ich Professor T Vamos und Professor M Thoma, chemalige Präsidenten der IFAC, für ihr ständiges Interesse an diesem Werk danken. Gerne würde ich auch Professor J F Coales, letzter Vorsitzender des IFAC Publications Managing Board, für seine Bemühung danken, dem Wörterbuch seine endgültige Gestalt zu geben.

Gerne danke ich allen aktiven Mitarbeitern meines technischen Kommittees die endgültige Englische Begriffssammlung vorbereitet zu haben und allen sprachlichen Herausgebern für die Übersetzung der Begriffe in ihre eigene Sprache.

Insbesondere möchte ich meiner Sekretärin, Fraa O Uibo, danken die sorgfaltig alle Arbeits unterlagen geschrieben hat. Ich bin auch meinen Kollegen des Instituts für Kybernetik und der technischen Universität Tallinn dankbar, die Text-Verarbeitungs-Systeme benutzten und mir halfen, Begriffe ins Russische zu übersetzen. Die Hilfe meines früheren Lehrers Professor H Sillamaa war unübertrefflich.

Ich wünsche meine aufrichtige Dankbarkeit jedem zum Ausdruck zubringen, dessen Mitwirkung die Veröffentlichung dieses Wörterbuchs ermöglicht hat. Ich selbst habe gern an dieser Arbeit teilgenommen.

Wie ich oben erwähnte, Kann ein Wörterbuch eines sich schnell ehtwickelnden Gebiets niemals vollständig sein. Seine Ergänzung ist vielmehr ein niemals endender Vorgang, der obsichtlich für kuize Zeit angehalten wird, um die Wörter-Sammlung in einem bestimmten Augenblick aufzunehmen und sie ohne Verzug vielen Menschen mit verschiedenen Sprachen zugänglich zu machen. So trägt es zu einem besseren verständnis in der Welt, wenigstens für die Steuerung und Regelung bei. Zugleich ist daran gedacht die Sprache zu beeinflussen, wo die bekettenden Begriffe erst beginnen sich zu entwickeln.

Ich hoffe, daß das nächste IFAC-Wörterbuch veröffentlicht sein wird bevor 20 Jahre vergehen und daß es auch Begriffe erfassen wird, die schon entstanden sind oder die gerade erst erscheinen, die wir leider nicht in der Lage sind, sie in das gegenwärtige Wörterbuch aufzunehmen.

Jegliche Kritischen Bemerkungen und Anregungen werden sehr geschätzt und sind an das IFAC-Sekretariat (Schloßplatz 12, A-2361 Laxenburg, Oesterreich) zu senden.

Ants Work
Vorsitzender des IFAC TC on TERMST

Professor Ants Work beendete, wie im Munchener Kongreß 1987 heschlossen, seine Amtszeit als Vorsitzender des IFAC Technischen Kommittees Begriffe und Normen (TC on TERMST). Seine hervorragende Arbeit als Vorsitzender des TC on TERMST ist in dieser zweiten Ausgabe des mehrsprachigen Wörterbuchs anzuerkennen. Die Nachtolge von Professor Work trat 1987 Professor Solheim an, der gewiß den Fortschritt der Herausgeber-Arbeit weitergeführt hätt, wäre ernicht durch seinen plötzlichen Tod nach dem Kongreß 1987 daran gehindert worden. Als stellvertretender Vorsitzender des TC on TERMST wurde ich dann aufgefordert mit Professor Otteblad den Vorsitz zu übernehmen. Wir beide verdanken viel den von Professor Work festgelegten Regeln und es war unser Hauptziel diese Reglen beizubehalten.

Die Entwicklung des Gebiets Regelungs und Steuerangs-Systeme ist niemals stehen, geblieben was sich auch in dieser zweiten Ausgabe bekundet, do sie mehr als 1500 Begriffe enthält. Die zunehmende gegenseitige Beeinflussung von Regelung und Steuerung durch die Rechner-Wissenschaft macht sich in der neuen Ausgabe bemerbar, aber nur solche Begriffe sind aufgenommen worden, die im Zusammenhang deutlich einen doppelten sinn haben.

Der Aufbau der zweiten Ausgabe unterscheidet sich duetlich von der unsprünglichen Ausgabe, in dem die englischen Begriffe alphabetisch aufgezählt sind und kein Versuch gemacht wurde, die Begriffe inhaltlich, wie in der vorherigen Ausgabe, zu ordnen. Diese Änderung wurde vorgenommen, um den Darstellungs methoden zu entsprechen wie sie in anderen Gebieten für mehr sprachige Wörterbücher üblich sind. Die Übersetzungen in die anderen sieben Sprachen sind in Spalten dem englischen Verzeichnis angegliedert. Em alphabetisches Verzeichnis der Begriffe für jede der anderen sieben Sprachen gibt die Verzeichnis-Nummer der entsprechenden englischen Übersetzung an. Es ist beabsichtigt, daß diese Darstellungs methode ein wirksameres und schnelleres Mittel der Übersetzung bietet.

Abschließend möchte ich meinen persönlichen Dank all denem zum Ausdruck bringen, die mit dem Projekt betaßtwaren, sowohl den IFAC-Mitgliedern für die Lieferung der Übersetzungen oder anderen Herausgeber-Materials, als auch den Mitarbeitern der Pergamon Press in Oxford, mit der ich während der letzten vier Jahre enge Verbindung hatte.

H A Prime

Многоязычный словарь ИФАК по автоматическому управлению

Предисловие

Прошло двадцать лет с тех пор как в 1967 г. был издан первый многоязычный словарь ИФАК по автоматическому управлению. Впервые предложение начать подготовку шестиязычного словаря было сделано во время второго конгресса ИФАК в Базеле в 1963 году и эта работа была поручена только что избранному комитету по терминологии. Ныне покойный проф. Д. Т. Бролбент был председателем этого комитета, а также главным редактором первого словаря ИФАК.

Главное внимание в первом издании было уделено упорядочению основных понятий автоматического управления, а также близких к этой области терминов. В словарь было включено ограниченное число терминов по элементам автоматики и некоторые термины из области вычислительной техники. Слова и выражения были упорядочены на английском языке по техническим областям и снабжены порядковыми номерами, а для всех языков составлены алфавитные указатели.

Дискуссии о необходимости переработки первого Словаря ИФАК начались в Техническом комитете ИФАК по терминологии и стандартам (ТК ТЕРМСТ) в 1972 году. Предложение тогдашнего председателя, доктора Х. Л. Мэйсона, добавить термины на японском языке было решено дополнить издание 1967 года японскими терминами и опубликовать этот семиязычный словарь к конгрессу ИФАК в Киото в 1981 г. Этот план был реализован вовремя издательством Pergamon Press благодаря очень эффективной работе проф. М. Масубучи, который отредактировал японские термины, представленные в двух азбуках. В это время проф. Бенеши подготовил словники ПРОДРОМУС I и II, из собранных членами ТК ТЕРМСТ новых терминов с предложением включить их в Словарь ИФАК. Начиная с конгресса ИФАК в Киото в 1981 г. автор этих строк был ответственным за сбор и отбор новых терминов на английском языке для включения их в обновленный словарь ИФАК, что нашло отражение в словниках ПРОДРОМУС III-VI.

Все, что мы достигли, — это результат совместной кропотливой работы активных членов ТК ТЕРМСТ. Нередко было трудно принять решение о включении термина — делая такой выбор, нельзя избавиться от субъективности. И такого рода работа никогда не может быть совершенной. Основными критериями при выборе новых терминов по автоматическому управлению для включения в словарь были следующие: термины не должны быть слишком общими или слишком специфическими (используемыми только узким кругом специалистов); они должны быть лаконичными; из области вычислительной техники может быть включен только бесспорно необходимый минимум терминов, иначе объем словаря оказался бы громадным. Кроме того, упорядочением терминологии в этой области наряду с другими международными организациями занимается ИФИП.

Во время конгресса ИФАК в Будапеште в 1984 г. со стороны ТК ТЕРМСТ было принято несколько очень важных, на наш взгляд, решений. Во-первых, новый Словарь ИФАК должен быть восьмиязычным (включая китайский). Во-вторых, термины следует упорядочить в алфавитном порядке. В-третьих, к терминам при необходимости нужно добавить краткие пояснения или определения на английском языке.

После конгресса в Будапеште подкомитет проанализировал издание 1981 г. и сделал заключение, что почти 100 терминов нет смысла включать в новый словарь. Сравнив в течение 1984 г. старый Словарь ИФАК (878 терминов) и словник ПРОДРОМУС VI (540 терминов) с изданием INSPEC THESAURUS 1983 г., проф. Г. А. Прайм нашел необходимым дополнить словарь многими новыми терминами. После обсуждения с членами ТК ТЕРМСТ из этих терминов было выбрано 142, как Дополнение к ПРОДРОМУС VI.

К концу 1985 года окончательный словник из английских терминов для нового многоязычного словаря ИФАК был готов. За перевод терминов на другие языки отвечали редакторы: Г. А. Феррате — испанский, Дж. Гоффман и П. Жилар — французский, М. Масубучи — японский, Р. Эткер — немецкий, Т. Раймонди — итальянский, А. Вырк — русский, Ву Ценг-Ксиан — китайский, и эта работа была завершена в октябре 1986 г.

Намерение снабдить некоторые термины краткими комментариями не нашло поддержки в Совете ИФАК по издательским делам, поскольку в недалеком будущем планируется издание в Pergamon Press многотомной Энциклопедии по управлению.

Теперь, после многолетней подготовительной работе новый многоязычный словарь ИФАК по автоматическому управлению наконец увидел свет, чему способствовали старания многих людей. Прежде всего я хочу поблагодарить бывших президентов ИФАК, профессоров Т. Вамоша и М. Тома за их постоянное внимание к нашей работе. Я приветствую бывшему председателю Совета ИФАК по издательским делам профессору Дж. Ф. Коулу за его ценные предложения при обсуждении окончательной формы словаря.

Приятно поблагодарить всех активных членов нашего технического комитета за подготовку основного словника из английских терминов и всех редакторов за перевод этих терминов на их родной язык.

Особенно я хочу отметить работу моего секретаря Ы. Уйбо, которая скрупулезно и аккуратно отпечатала почти все рабочие материалы. Я благодарен также моим коллегам из Института кибернетики и Таллиннского технического университета за предоставленную возможность использовать систему обработки текстов и за помощь при переводе терминов на русский язык. Помощь моего бывшего учителя профессора Х. Силламаа была незаменимой.

Свою искреннюю благодарность я адресую всем, чье участие сделало возможным издание этого словаря. Для меня лично участвовать в этой работе было удовольствием.

Как я уже отметил, словарь терминов быстро развивающейся научной области никогда не может быть совершенным. Составление подобных изданий по существу является бесконечным процессом, который приходится намеренно останавливать для фиксации состояния терминологии в конкретный момент времени, чтобы без сильного запаздывания сделать ее доступной многим людям, говорящим на разных языках, и тем самым улучшить взаимопонимание в мире, в данном случае — в области автоматического управления. Одновременно достигается и другая цель — повлиять на развитие тех языков, где соответствующая терминология только-только начинает развиваться.

Я уверен, что следующий словарь ИФАК будет издан раньше, чем пройдет еще 20 лет, и в него будут включени те термины, которые уже появились или только появляются и которых мы не в состоянии были включить в настоящий словарь.

Все критические замечания и предложения будут приняты нами с интересом и благодарностью и могут быть адресованы в секретариат ИФАК (Schlossplatz 12, А-2361 Laxenburg, Austria).

Антс Вырк,
Председатель ТК ТЕРМСТ ИФАК

xiii

Председатель Технического Комитета ИФАК по терминологии и стандартам Антс Вырк занимал эту должность до конгресса ИФАК в Мюнхене в 1987 г. Значимость его работы в этом качестве получит несомненное признание в связи с опубликованием этого второго издания Многоязычного словаря. Преемником А. Вырка в ТК ТЕРМСТ стал проф. Солхайм, который, несомненно, продолжил бы редакционную работу над словарем, если бы так неожиданно не скончался вскоре после конгресса в 1987 г. Как заместитель председателя ТК ТЕРМСТ мне предложили принять председательствование, и проф. Оттэлод стал моим заместителем. Мы оба по достоинству оценили уровень требований, установленный А. Вырком, и нашей основной целью было соблюдение этого уровня.

Развитие областей управления и системотехники всегда отмечалось динамичностью, чем свидетельствует и тот факт, что в настоящем издании содержится уже более 1500 терминов. Растущая взаимосвязь между наукой об управлении и вычислительной техникой проявляется и в новом издании словаря, хотя сюда включены только те термины, которые в каждой из этих областей имеют свое специфическое толкование.

По форме второе издание существенно отличается от первого в том смысле, что английские термины снабжены порядковыми номерами, а последовательность их чисто алфавитная и не сделано никаких попыток группировать термины по смыслу, как было в предыдущем издании. Такое изменение сделано с учетом того, что подобные принципы представления терминов используются в аналогичных многоязычных словарях по другим техническим дисциплинам. Переводы на остальные семь языков приведены в столбце, примыкающем к английскому термину. Для каждого из семи языков имеются алфавитные указатели с порядковыми номерами соответствующих английских терминов. Это позволяет более эффективно и быстро пользоваться словарем для перевода.

В заключение мне хотелось бы выразить свою личную благодарность всем, кто был связан с выходом в свет этого издания, и особенно, членам ТК ТЕРМСТ, которые подготовили переводы терминов и другие материалы, и сотрудникам Pergamon Press в Оксфорде, с которыми я был тесно связан последние четыре года.

Г. А. Прайм

Dizionario Multilingue IFAC di Controlli automatici

Prefazione

Molti anni sono trascorsi dalla prima pubblicazione, nel 1967, del Dizionario Multilingue IFAC sul lessico dei Controlli automatici. Fu durante il 2° Congresso IFAC, tenutosi a Basilea nel 1963, che venne avanzata la prima proposta di avviare le procedure per la pubblicazione di questo dizionario esalingue; e il mandato fu assunto da un Comitato eletto ex novo. Lo scomparso Prof. D T Broadbent, che fu anche il Curatore generale del primo Dizionario IFAC, funse da Presidente del Comitato.

L'attenzione fu soprattutto rivolta ai concetti fondamentali dei controlli automatici e alla corrispondente terminologia, ma s'introdusse anche un ristretto numero di termini fortemente connessi e relativi ai componenti, nonché alcuni termini mutuati dalla tecnologia informatica.

Il primo Dizionario consisteva di un Vocabolario, nel quale parole ed espressioni erano ordinate per argomento, e sei Indici alfabetici, che consentivano di localizzare un dato termine nel Vocabolario.

La discussione sulla necessità di rivedere il primo Dizionario IFAC ebbe inizio nell'ambito del Comitato tecnico IFAC su Terminologia e Standards (TERMST) nel 1972. Il suggerimento di aggiungere i termini giapponesi, avanzato dal Presidente uscente Dr. H L Mason, fu calorosamente appoggiato e fatto proprio dal nuovo Presidente Professor J Benes. Nel 1980, si decise di aggiungere i termini giapponesi all'edizione originale del 1967 e di pubblicare il Dizionario come glossario eptalingue presso la PERGAMON PRESS, in tempo per il Congresso IFAC di Kyoto del 1981. Questo progetto fu puntualmente concluso grazie all'efficientissimo lavoro del Professor M Masubuchi, curatore della versione giapponese (sia con caratteri Kanji che Kana).

Intanto, il Professor Benes curò PRODROMUS I e II, gli elenchi di nuovi termini sui controlli automatici, raccolti dai membri di TERMST per promuoverne l'inclusione nel nuovo Dizionario IFAC.

A partire dal Congresso di Kyoto del 1981, l'attuale Presidente del Comitato tecnico TERMST si è assunto il compito di raccogliere e selezionare nuovi termini in inglese per l'aggiornamento del Dizionario IFAC, termini che vennero ordinati nei testi del lavoro PRODROMUS da III a VI.

Quello che abbiamo conseguito è opera prevalentemente collettiva dei membri attivi di TERMST. Qualche volta è stato difficile decidere se includere un termine oppure ometterlo. L'attuale selezione può essere molto soggettiva, ma un lavoro come questo non può mai considerarsi finito. L'idea di fondo nel decidere l'inclusione di nuovi termini sui controlli automatici è stata la seguente: i termini non devono essere né troppo generali né troppo specifici (usati solo in certe applicazioni); non devono essere troppo lunghi, e non dev'esserci che un minimo di termini desunti dalla tecnologia informatica, pena la crescita esorbitante del numero di termini nel Dizionario. Per altro, quest'ultimo campo terminologico è coperto dall'IFIP, fra altri organismi internazionali.

Durante il Congresso IFAC di Budapest, nel 1984, il Comitato tecnico TERMST prese diverse decisioni importanti (dal nostro punto di vista). Innanzitutto, il nuovo Dizionario Multilingue IFAC avrebbe contemplato otto lingue (compreso il cinese). Secondariamente, tutti i termini sarebbero stati posti in ordine alfabetico. Terzo, i termini sarebbero stati corredati di brevi chiose o definizioni in inglese.

Dopo il Congresso di Budapest, un sottocomitato riprese in esame il Dizionario IFAC (edizione 1981), e a seguito di quel lavoro quasi 100 termini vennero ritenuti superati nel nuovo Dizionario. Durante il 1984, il Professor H A Prime comparò il vecchio Dizionario IFAC (878 termini) e il PRODROMUS VI (540 termini) con l'INSPEC THESAURUS (edizione 1983) trovando molgi termini

aggiuntivi. Dopo una discussione con i membri del Comitato, venne comilato un elenco di 142 termini quale AGGIUNTA a PRODROMUS VI.

Entro la fine del 1985, l'insieme definitivo di termini in lingua inglese per il nuovo Dizionario Multilingue IFAC era pronto. La traduzione dei termini nelle altre lingue venne assicurata dai curatori linguistici (G A Ferratè per lo spagnolo, J Hoffmann e P Gilard per il francese, M Masubuchi per il giapponese, R Oetker per il tedesco, T Raimondi per l'italiano, A Work per il russo, Wu Zeng-Quian per il cinese); questo compito era terminato nell'ottobre del 1986.

L'intenzione di aggiungere brevi chiose o commenti a vari termini venne abbandonata, su suggerimento dell'IFAC Publications Managing Board, in vista della pubblicazione già programmata di un'Enciclopedia del controllo, in più volumi, presso Pergamon.

Ora, dopo molti anni di lavoro preparatorio, il nuovo Dizionario Multilingue IFAC di Controlli Automatici viene pubblicato. Io vorrei prima di tutto ringraziare il Professor T Vamos e il Professor M Thoma, già Presidenti dell'IFAC, per il loro continuo interesse in questo lavoro. Vorrei ringraziare il Professor J F Coales, ex Presidente dell'IFAC Publications Managing Board, per la parte da lui avuta nel dare al Dizionario la sua forma definitiva.

Vorrei anche ringraziare tutti i membri attivi del mio Comitato tecnico per la messa a punto dell'insieme finale di termini in inglese, e tutti i curatori linguistici per la traduzione dei termini nella loro lingua.

Vorrei ringraziare in particolare la mia segretaria, Sig.ra O Uibo, che ha battuto diligentemente a macchina tutto il materiale di lavoro. Sono anche grato ai miei colleghi dell'Istituto di Cibernetica e del Politecnico di Tallinn che si sono avvalsi dei sistemi per l'elaborazione dei testi e mi hanno aiuto nella traduzione in russo. Il sostegno del mio maestro di un tempo, Professor H Sillamaa, è stato indispensabile.

Voglio esprimere la mia sincera gratitudine a tutti coloro che, con il loro contributo, hanno reso possibile la pubblicazione di questo Dizionario. Per me è stato un piacere avere una parte in questo lavoro.

Come ho ricordato precedentemente, un dizionario riguardante un campo in rapida evoluzione non può mai considerarsi finito. La sua comilazione è piuttosto un processo senza fine deliberatamente interrotto, a un tratto, per registrare il lessico di un determinato momento, rendendolo immediatamente accessibile a molte persone di madrelingua diversa, e contribuire così ad una miglior comprensione nel mondo (almeno) dei controlli automatici. Con ciò s'intende, nel medesimo tempo, esercitare un'influenza sui linguaggi, laddove la rispettiva terminologia stia appena cominciando a svilupparsi.

Confido che il prossimo Dizionario IFAC venga pubblicato prima che trascorrano altri 20 anni, e che esso contenga anche quei termini già creati, o in procinto di affermarsi, che non siamo stati in grado d'includere nel Dizionario attuale.

Ogni suggerimento o commento critico, eventualmente indirizzato a IFAC Secretariat (Schlossplatz 12, A-2361 Laxenburg, Austria), sarà estremamente gradito.

Ants Work
Presidente del Comitato tecnico TERMST dell'IFAC

Il Professor Ants Work terminò il suo mandato di Presidente del Comitato IFAC denominato TERMST alla conclusione del Congresso di Monaco nel 1987. Il suo eminente lavoro quale Presidente di TERMST sarà riconoscibile nella pubblicazione di questa seconda edizione del Dizionario multilingue. Il Professor Solheim, che succedette nel 1987 al Professor Work, avrebbe certamente continuato ad assicurare l'avanzamento del lavoro editoriale, non fosse stato per la sua intempestiva scomparsa, a così breve distanza dal Congresso del 1987. Quale Vice-Presidente de TERMST, fui allora invitato a subentrare alla Presidenza e il Professor Otteblad divenne Vice-Presidente. Entrambi dobbiamo molto ai livelli di prestazione stabiliti dal Professor Work; il nostro obbiettivo primario è stato infatti il mantenimento di quei livelli.

Lo sviluppo delle discipline dei sistemi e del controllo non è mai divenuto statico, e ciò è manifesto in questa seconda edizione per il fatto ch'essa include più di 1500 termini. Anche la crescente interazione fra automatica e informatica è visibile nella nuova edizione, ma solo quei termini che chiaramente attengono al duplice contesto sono stati inclusi.

La struttura della seconda edizione è pure notevolmente diversa da quella dell'edizione originale in quanto l'ordine di elencazione dei termini in inglese è interamente alfabetico e non è stato fatto alcun tentativo di raggruppare termini nell'ambito di un particolare contesto, come nella precedente edizione. Questa modifica è stata introdotta al fine di conformarsi ai metodi di presentazione adottati in altri glossari multilingue in altri settori disciplinari. Le traduzioni nelle altre sette lingue sono quindi ordinate in colonne adiacenti a quella dell'elenco in inglese. Un indice alfabetico dei termini in ciascuna delle altre sette lingue ha lo scopo di fornire il numero d'ordine dell'equivalente traduzione in inglese. Si è ritenuto che questo metodo di presentazione fornisse uno strumento di traduzione più rapido ed efficace.

In conclusione, verrei esprimere i miei personali ringraziamenti a tutti coloro che hanno partecipato al progetto, vuoi fornendo traduzioni o altro materiale editoriale come membri dell'IFAC, vuoi come membri della Pergamon Press di Oxford, coi quali sono stato in stretto collegamento durante i quattro anni trascorsi.

H. A. Prime

Diccionario Multilingüe de Automática de IFAC

Prólogo

Han pasado veinte años desde que en 1967 se publicó el Primer Diccionario multilingüe de IFAC sobre terminología de Automática. Fue durante el Segundo Congreso Mundial de IFAC, en 1963 en Basilea, cuando se planteó la primera propuesta para iniciar los preparativos de la edición de este diccionario en seis idiomas, responsabilizándose un Comité recién elegido. El Profesor D T Broadbent ejerció como Presidente de dicho Comité, siendo también Director General de la publicación del Primer Diccionario de IFAC.

Se puso especial énfasis en los conceptos básicos de la automática y la correspondiente terminología. Se incluyó un número reducido de términos estrechamente relacionados con componentes, así como unas pocas palabras relativas a la tecnología de los computadores.

El Primer Diccionario consistió en un Vocabulario, en el cual se clasificaron las palabras y expresiones por ttemas técnicos, y seis Indices alfabéticos que posibilitaban la localización de una determinada palabra en el Vocabulario.

Los debates sobre la necesidad de revisar este Primer Diccionario de IFAC se iniciaron en 1972, en el Comité Técnico sobre terminología y normas (TC on TERMST). Se apoyó formalmente la propuesta del Primer Presidente Dr. H L Mason de incluir términos japoneses, siendo adoptada por el siguiente Presidente Profesor J Benes. En 1980, se decidió incluir la terminolgia japonesa en la Edición original de 1967 y publicar el Diccionario con un glosario en siete idiomas, por PERGAMON PRESS, a tiempo para el Congreso de IFAC en Kyoto en 1981. Este proyecto se completó a su debido tiempo gracias al muy eficiente trabajo del Profesor M Masubuchi, Director de la parte japonesa (utilizando tanto caracteres Kanji como Kana).

Entre tanto, el Profesor Benes dirigió la publicación de PRODROMUS I y II, listas de nuevos términos sobre la automática, recogidas por los miembros del TC on TERMST y propuestas para su inclusión en el nuevo Diccionario de IFAC.

Desde el Congreso de Kyoto en 1981, el actual Presidente del TC on TERMST se ha responsabilizado de reunir y seleccionar nuevos términos en inglés, para actualizar el Diccionario de IFAC, apareciendo catalogados en los documentos de trabajo PRODROMUS III al VI.

Lo que se ha conseguido es el resultado de una tarea fundamentalmente colectiva de los miembros activos del TC on TERMST. A veces, ha habido dificultades para decidir la inclusión u omisión de un término. La decisión adoptada ha podido ser muy subjetiva. Pero un trabajo de este tipo nunca puede ser completo. Las ideas clave para decidir la inclusión de nuevos términos sobre la Automática han sido las siguientes; los términos no debían ser ni demasiado generales ni demasiado específicos (sólo usados en ciertas aplicaciones), no debían ser demasiado largos; debía haber sólo un mínimo de palabras referidas a terminología de computadores, ya que de lo contrario el número de palabras del Diccionario crecería enormemente. Además, la terminología de esta última área quedaba cubierta por IFIP, entre otros organismos internacionales.

Durante el Congreso de IFAC de Budapest, en 1984, se tomaron varias decisiones importantes (desde nuestro punto de vista) por parte del TC on TERMST. En primer lugar, el nuevo Diccionario Multilingüe de IFAC tendría ocho idiomas (incluyendo chino).

En segundo lugar, las palabras se ordenarían alfabéticamente. En tercer lugar, cuando fuera conveniente los términos irían acompañados de cortas explicaciones ó definiciones en inglés.

Después del Congreso de Budapest, un Subcomité revisó el Diccionario de IFAC (edición 1981) y, como resultado de este trabajo, casi 100 palabras se consideraron innecesarias para incluir en el nuevo Diccionario. Durante 1984, el Profesor H A Prime comparó el antiguo Diccionario de IFAC (878 términos) y PRODROMUS VI (540 términos) con el INSPEC TESAURUS (edición 1983),

encontrando muchos términos adicionales. Tras ser debatido con los miembros del TC, se incluyó una lista de 142 términos como Adenda a PRODROMUS VI.

El conjunto final de términos en inglés para el nuevo Diccionario Multilingüe de IFAC estuvo listo hacia finales de 1985. La traducción de los términos a otros idiomas fue responsabilidad de los Directores de la publicación en cada idioma (G A Ferraté para español, J Hoffman y P Gilard para francés, M Masubuchi para japonés, R Oetker para alemán, T Raimondi para italiano, A Work para ruso y Wu Zeng-Qian para chino). Este trabajo se completó en Octubre de 1986.

Se abandonó la intención de añadir explicaciones cortas o comentarios a algunos términos, debido a la sugerencia del proyectos futuros de publicar una Enciclopedia multivolumen de Automática por Pergamon.

Ahora, tras muchos años de trabajo preparatoria, se publica el nuevo Diccionario Multilingüe de Automática de IFAC. Quisiera, en primer lugar, dar las gracias a los Profesores T. Vamos y M. Thoma, entiguos Presidentes de IFAC, por su permanente interes en esta empresa. Quisiera, también, agradecer al Profesor J F Coales, antiguo Presidente del Equipo Directivo de publicaciones de IFAC, por su contribución en dar la forma definitive al Diccionario.

Quisiera también dar las gracias a todos los miembros activos del Comité Técnico por la preparación del conjunto final de términos en inglés y a los responsables de los diferentes idiomas por traducir la terminología en su lengua materna.

En especial, quiero agradecer a mi secretaria, la Sra. O Uibo, quien con sumo cuidado mecanografió casi todos los documentos de trabajo. Le estoy también especialmente agradecido a mis compañeros del Instituto de Cibernética y de la Universidad Técnica de Tallin que utilizaron los (procesadores de textos) sistemas de tratamiento de textos y me ayudaron en la traducción de los términos al ruso. La ayuda de mi primer maestro, el Profesor H Sillamaa, ha sido fundamental.

Quiero expresar mi sincero agradeoimiento a todos aquellos cuya ayuda ha hecho posible la publicación de este Diccionario. Para mí, ha sido un placer participar en este trabajo.

Como comenté previamente, un diccionarie en un campo de desarrollo tan rápido nunca puede ser completo. La tarea de recopilación es un proceso que no tiene fin; y, por tanto, se tiene que detener intencionadamente en un determinado momento, a fin de recopilar la terminolgía existente y, sin retraso, hacerla accesible a mucha gente con diferentes idiomas, contribuyendo a un mejor entendimiento (al menos) en el mundo de la automática. Ello significa, al mismo tiempo, influir sobre los distintos idiomas en el desarrollo de las respectivas terminologías cuando se están empezendo a acuñar.

Confío en que el siguiente Diccionario de IFAC se publicará antes de que trascurran otros 20 años y que incluirá aquellos términos ya creados o a punto de aparecer, y que nosotros no hemos podido incluir en el presente Diccionario.

Se agradecerá enormemente cualquier crítica o sugerencia, pudiendo dirigirlas al Secretariado de IFAC (Schlossplatz 12, A-2361 Laxenburg, Austria).

Ants Work
Presidente del TC on TERMST de IFAC

Postdata al Preambulo

El Profesor Ants Work completó su periodo como Presidente del Comité Técnico de IFAC sobre Terminología (TERMS) al final del Congreso de Munich, en 1987. El extraordinario trabajo llevado a cabo como Presidente del Comité TERMS queda de manifiesto con la publicación de esta sequanda edición del Diccionario Multilingüe. El Profesor Work fue sucedido en 1987 por el Profesor Soldheim quien sin duda habría continuado coneste trabajo editorial a no ser por su imprevistat muerte, poco después del Congreso de 1987. Como Vice-Presidente de TERMS, fui invitado a ocpar la Presidencia, pasando a ser Vice-Presidente el Profesor Ottblad. Las normas establecides por el Profesor Work fueron de enorme utilidad, siendo uno de nuestros objetivos básicos el mantenerlas a lo largo de toda la edición.

El desarrollo de las disciplinas de control y sistemas no ha sido nunca estático, como lo prueba el hecho de que estat Segunda Edición incluye mas de 1500 términos. También es evidente, en esta nueva edición, la creciente interacción entre el control y las ciencias de la computación, aunque solamente se han incluido aquellos términos que claramente tienen una doble interpretación.

El formato de la Segunda Edición también es significativamente diferente al de la edición original, en el sentido de que el listado numérico para log términos en inglés es puramente alfabético y no se ha tratodo de agrupar los términos por temas, como en la edición previa. Este cambio se ha hecho para adecuarnos a los métodos de presentación adoptados en glosarios multilingues de otras disciplinas técnicas. La traducción a los otros siete idiomas estat listada en columnas advyacentes a la de Inglés. Se incluye un Índice, alfabético de los términos en cada uno de los siete idiomas, proporcionando el número de llistado de la correspondiente traducción en Inglés. Se entiende que esta forma de presentación proporciona una búsqueda más rápida y efectiva para la traducción.

Para terminar, me gustaría expressar mi agradecimiento personal a todos los que han participado en el proyecto, ya sean miembros de IFAC que han proporcionado las traducciones u otro material editorial, ya sean personal de Pergamon Press, de Oxford, con los que he mantenido una estrecha relación durante los pasados cuatro años.

自動制御に関するＩＦＡＣ多国語辞典

序言

　1967年に自動制御用語についての最初のＩＦＡＣ多国語辞典が発行されてから20年経過した。この６ヶ国語辞典を編集する準備の初めての提案がなされた時は、1963年に Basle で開催された第２回ＩＦＡＣ大会開期中であり、新しく選ばれた委員会が責任を持った。故 D.T.Broadbent 教授が委員会の委員長となり、また、最初のＩＦＡＣ辞典の総編集者でもあった。

　主力は自動制御の基礎概念とそれに関した用語におかれた。要素については限られた数の密接に関連ある用語および計算機技術の分野からの若干の用語も取り入れられた。

　最初の辞典は、言葉と表現を技術的トピックスに従って配慮した語彙と、語彙中の用語の位置を示す６組のアルファベット順の索引から成っていた。

　最初のＩＦＡＣ辞典の改訂の必要性についての討論が、1972年の「用語と標準に関するＩＦＡＣ専門委員会（TC on TERMST）」にて始まった。前委員長 Dr.H.L.Mason による日本語用語を追加しようという提案が熱心に支持され次期委員長 J.Beneš 教授によって採用された。日本語用語を1967年の初版に追加して７ヶ国語辞典とし、1981年のＩＦＡＣ京都大会に間に合うように PERGAMON 社から出版することが1980年に決定された。この計画は日本（漢字とカナ文字を使って）の編集委員である増淵正美教授の非常に効果的な骨折りによって予定期日に完成された。

　一方、Beneš 教授は「用語と標準に関する専門委員会」の委員から自動制御の新用語のリストを集めてＰＲＯＤＲＯＭＵＳ　Ｉ　および　II　を編集し、新ＩＦＡＣ辞典に入れるよう提案した。1981年の京都大会以降、「用語と標準に関する専門委員会」の現委員長は、増訂ＩＦＡＣ辞典のための英語の新用語の収集と選定に責任を取り続け、ＰＲＯＤＲＯＭＵＳ　III　VI　にまとめた。

　我々の到達したものは、「用語と標準に関する専門委員会」の活発なメンバーの主として共同作業の成果である。時には一つの用語を入れるべきか除くべきかを決定することが困難だったこともあった。現在の選択は非常に主観的であるかもしれない。しかし、このような作業は決して完了することはない。自動制御に関する新用語を包含させるための選定中の基本理念は以下のようなものであった：用語は一般すぎてはならない、また、特殊（あるところに使われるだけの）すぎてもいけない；長すぎてはならない；また、計算機技術用語は最少にすべきである、さもなければ辞典中の用語数は限りなく増大してしまうであろう。なお、後者の分野の用語は他の国際的団体であるＩＦＩＰに包含される。

　1984年のＩＦＡＣブタペスト大会中に我々の観点からの幾つかの重要な決定が「用語と標準に関する専門委員会」によってなされた。第一に、新ＩＦＡＣ多国語辞典は８ヶ国語（中国語を含んで）を取り入れること。第二に、すべての用語はアルファベット順とすること。第三に、できることなら用語は短い説明か定義を英語でつけること。

　ブタペスト大会後、小委員会がＩＦＡＣ辞典（1981年版）をチェックし、その作業の結果、ほぼ100の用語が新辞典には不要であると考えられた。1984年中に H.A.Prime 教授は旧ＩＦＡＣ辞典（878用語）と PRODROMUS VI（540用語）とを INSPEC 語彙集（1983年版）と比較し、多数の追

加用語を見出した。用語委員会メンバーとの討論後、142用語のリストが PRODROMUS VI への増補として作られた。

　1985年末までに新ＩＦＡＣ辞典のための英語用語の最終版の準備ができた。用語の他国語への翻訳は国際語編集委員（G.A.Ferrate スペイン語、J.Hoffmann および P.Gilard フランス語、増淵 日本語、R.Oetker ドイツ語、T.Raimondi イタリー語、A.Wörk ロシヤ語、Wu Zeng-Qian 中国語）の責任であり、この仕事は1986年10月に完了した。

　幾つかの用語にある短い説明か注釈を付ける意図は、多分冊の Pergamon 制御百科辞典発行の将来計画を持つＩＦＡＣ出版管理委員会の提案によって放棄した。

　さて、多年にわたる準備作業を経て自動制御に関する新ＩＦＡＣ多国語辞典が出版されることになった。私は先ず T.Vamos 教授とＩＦＡＣ前会長 M.Thoma 教授に対し、この仕事に絶えず興味を持って頂いたことを深謝したい。また、ＩＦＡＣ出版管理委員会の元委員長の J.F.Coales 教授に対し、本辞典に最終的な具体化をされたことを感謝したい。

　さらに、用語の英語での最終設定の準備された本専門委員会の活発な全委員とその用語を各母国語に翻訳された全国際語委員に感謝したい。

　また、ほとんどすべての資料を注意深くタイプした私の秘書 O.Uibo 婦人に感謝したい。それから資料作成の便宜やロシヤ語への翻訳に助力されたサイバネティクス研究所とタリン工科大学の同僚に謝意を表する。私の以前の教師 H.Sillama 教授の助力も欠かすことができないものであった。

　本辞典の出版を可能ならしめたすべての方々に対し私の深い謝意を表したい。私にとってこの仕事に参加できたのは嬉しいことであった。

　前述のように急速に発展している分野の辞典は決して完成され得ない。一つを編集するということは、ある特定の瞬間の用語を記録し、即刻、異なった母国語を持つ多くの方々に伝えられるようにするために、漸時意図的に停めたいという、むしろ決して終わることの無い過程なのであり、こうすることによって自動制御の世界（少なくとも）におけるより良い理解に貢献するのである。同時に一つ一つの用語が正に発展し始めている言葉に影響を与えるという意味もある。

　私は次のＩＦＡＣ辞典はこれから20年経過しない内に出版されるであろうと信ずるが、さらに既に創生されている用語、あるいは現在の辞典に含ませることはできなかったが丁度出現しようとしている用語をも含むであろうと信ずる。

　御批判と御提案はすべて大いに歓迎する。ＩＦＡＣ事務局（schlossplatz 12, A-2361 Laxenburg,Austria）までに一報されたい。

　　　　　　　Ants Wörk
　　　　　　　委員長、用語と標準に関するＩＦＡＣ専門委員会

序文に対するあとがき

　1987年のミュンヘン大会の終了に当たり、Ants Wõrk 教授はＩＦＡＣの「用語と標準に関する専門委員会」の委員長としての任期を完遂した。彼の同委員会の委員長としての顕著な業績はこの多国語辞典の第2版の出版に認められる。 Wõrk 教授の後を1987年に Solheim 教授が継いだが、同年の大会直後の突然の死去がなければ編集の仕事の進捗を維持したことは確実であった。

　同委員会の副委員長として私は招待されて委員長の任に着き、Otteblad 教授が副委員長となった。我々二人は共に Wõrk 教授によって作られた基準に負うところが多く、これらの基準を維持することが我々の主な目的であった。

　制御とシステムの学問分野の進歩は決して静的ではなく、この第2版では1500以上の用語を包含しているという事実によってこの事は明白である。制御と計算科学との間の相互関係の増加も新版では明らかであるが、明瞭な二重関係を持つ用語のみ包含された。

　第2版の体裁は元の版の体裁に対しまた著しく異なっているが、それは配列が英語用語に対して全くアルファベット順であるが、特定の関係内にあるグループ用語については以前の版におけるように何らの試みも行ってない。この変動は他の専門分野の多国語用語辞典に採用されている表示法に従うために行ったのである。他の7つの国語への翻訳は英語の配列に接近して縦行に並べた。他の7ヶ国語のそれぞれの用語のアルファベット順の索引は同等の英語翻訳の配列数を与えるように含ませた。この表記法は翻訳がより有効により早くできるように企画されている。

　終わりに当たり，ＩＦＡＣのメンバーとして翻訳や他の編集資料を提供された方々、あるいは、過去4年間にわたって私と密接に仕事をしてきた Oxford の Pergamon 出版社の社員としてこの計画に没頭してきたすべての方々に、私は個人的な謝意を表したい。

H.A.Prime

IFAC 多种语言自动控制词典

序言

自从 IFAC 多种语言自动控制词典第一版在 1967 年同世以来，二十年已经过去了。那是在 1963 年 Basle 召开的第二次 IFAC 大会上，首先提出建议要准备出版一部具有六种语言的自动控制词典，并由新选举的委员会负责此项工作。D.T.Broadbent 教授（已故）当时任委员会主席，同时也担任 IFAC 第一本词典的总编辑。

词典重点编辑自动控制基本概念及其有关的术语，同时也包括有限的与自控元件有密切关系的术语和少量计算机技术方面的术语。词典第一版中的词汇是按专业次序编排的，并附有按字母顺序排列的六种语言的索引，以便对具体术语的检索。

在 1972 年 TERMST 委员会上，对第一版是否需要进行修改曾进行了讨论。前主席 H.L.Mason 博士建议把日文增加进去，并得到了继任主席 J.Benes 教授的支持。

1980 年正式决定把日文增加到 1967 年的版本中，并作为具有七种语言的词典予定在 1981 年 IFAC 京都（Kyoto）大会召开之际，由 PERGAMON 出版社出版。由于负责日文编辑的 M.Masubuchi 教授（采用了汉字和假名）非常有效的工作，这项计划得以及时完成。

在此期间，Benes 教授编辑了 PRODROMUS Ⅰ 和 Ⅱ，这些自动控制方面的新词条是由 TERMST 委员会各委员搜集并建议要增加到新的词典中。

1981 年京都大会后，为了不断更新 IFAC 词典，TERMST 委员会主席一直负责搜集选择新的英语词条，并汇集在 PRODROMUS Ⅲ ～ Ⅳ 中。

我们现在已经做到的都是 TERMST 委员会各成员积极的集体工作成果。在工作中对某一个词条的取舍，有时会感到非常困难。当然对目前已收入的词条也可能带有一定的主观性。但是这类性质的工作永远不会是完美无缺的。我们选择自动控制新词条的基本思想是：词条不应过于一般化，也不应过窄（不应仅适于个别应用）；词条不应过长；计算机技术方面的词条应限制在最少量，否则词典的词条量就会过大，况且这类词条 IFIP 等其它国际组织也均已收集。

1984 年 IFAC 布达佩斯（Budapest）大会期间，TERMST 委员会作了几项重要决定（根据我们的看法）：(1)新的 IFAC 多种语言词典应包含八种语言（把中文包括在内）。(2)全部词条应按字母顺序排列。(3)在适当的地方，词条应附有简短的英文注释或定义。

布达佩斯大会后，有一个专门小组重新校核了 IFAC 词典（1981 年版），校核结果认为有 100 个词条应从将来新的词典中删去。在 1984 年期间，H.A.Prime 教授对旧的 IFAC 词典（含 878 个词条）和 PRODROMUS Ⅳ（540 个词条）与 INSPEC

THESAURUS（1983 年版）进行了对比，发现了许多另外的词条。通过委员会成员的讨论，有 142 条增加到 PRODROMUS Ⅵ 中。

在 1985 年年末，IFAC 新的多种语言词典业已准备就绪。对英文以外其它几种语言的翻译分别由有关语言编辑负责（G.A.Ferrate 负责西班牙文，J.Hoffmann 和 P.Gilard 负责法文，M.Masubuchi 负责日文，R.Oetker 负责德文，T.Raimondi 负责意大利文，A.Work 负责俄文，Wu Zeng-Qian 和 Wang Zi-Ping 负责中文）。这项工作在 1986 年 10 月完成。

由于 IFAC 出版管理局在将来有出版 Pergamon 自动控制百科全书的计划，故取消了原来要增加词条注释的打算。

现在通过多年的准备工作，新的 IFAC 多种语言自动控制词典终于出版了。首先我想对 IFAC 前主席 T.Vamos 教授及 M.Thoma 教授对这项工作的大力支持表示感谢；对出版管理局前主席 J.F.Coales 教授关于词典最终校样所做的贡献表示感谢。

我还想向我们委员会各位委员的积极工作，对英文词条最终稿的准备以及各位语言编辑把英文词条译成各自的本国语言所做的工作表示感谢。

我要特别感谢的是我的秘书 O.Uibo 夫人，她细心地打印了几乎全部的文稿。还要感谢控制论学院（Institute of Cybernetics）和塔林技术大学（Tallinn Technical University）中我的一些同事，他们利用文本处理系统，帮助我把词条译为俄文。还要感谢我原先的导师 H.Sillamaa 教授，他的帮助也是不可少的。

我要向帮助这本词典得以出版的每一个人表示衷心的感谢。我本人得以参加这项工作也深感荣幸。

正如前面所说，一个发展极快领域的词典永远不可能完美无缺。更确切地说，一部词典的编辑是一个永无休止的过程。人们不可能停下一会儿，以便无延迟地记录下某一特定时刻的术语，以供不同语言的人进行查阅，从而有助于（至少）自动控制界能有较好的了解。同时这样做也意味着要对即将产生术语词条的那些语言施加影响。

我深信在未来二十年里，另一部 IFAC 词典将会出版。其中将包括现在已经产生或即将产生的那些词条，而这些词条我们却未能收入到目前这本词典中。

非常欢迎各种批评和建议，并请将它们寄至 IFAC 秘书处（Schlossplatz 12，A-2361 Laxenburg，Austria）。

IFAC TERMST 委员会主席　Ants　Work

序的附言

1987 年 *MUNICH* 大会闭幕时，*Ante Work* 教授即圆满地结束了他作为 *IFAC TERMST* 委员会主席的任期。他作为 *TERMST* 委员会主席的杰出贡献可以从多种语言词典第二版的出版得到公认。

在 1987 年接任 *Work* 教授主席职务的是 *Solhem* 教授，他（*Solhem* 教授）在 1987 年大会后不久即不幸逝世。如果他仍健在，毫无疑问地他会继续保证这项编辑工作的进展。当时我作为 *TERMST* 委员会的副主席被邀请接任了主席职务。*Otteblad* 教授接任了副主席职务。我们俩人依靠 *Work* 教授制定的编辑工作标准，并把坚持这些标准做为我们工作的主要目标。

控制与系统学科的发展从未停止过，这一点从第二版中所包含的 1500 个词条这一事实就可明显地看出。此外控制科学与计算科学之间不断增长的交互影响，从新版本内容看也是明显的。不过所收入的词条只是那些与两方面都有意义的词条。

第二版的编排方式与第一版也有很大不同。新版中用数字表示的英文词条顺序是完全按照字母表顺序排列，不再是象第一版那样按专业分类顺序排列。这种改变是为了与其它学科的多种语言词典的表示方法相一致。另外七种语言的词条列在对应英语词条的旁边。新版本还包括另外七种语言词条按字母顺序排列的索引，利用此索引可找到相应英文词条的排列数码。采用这种表示方法，目的在于提供更有效、更快的词条翻译。

最后我想向每一位对此项工作作出贡献的人表示感谢，包括那些提供各种翻译或编辑资料的 *IFAC* 成员以及在牛津 *Pergamon* 出版社的各成员，四年来我同他们进行了密切的合作。

H.A.Prime

EIGHT-LANGUAGE GLOSSARY

#	English	français	Deutsch	русский	italiano	español	日本語	中文
1	a.c. tacho-generator	génératrice f tachymétrique à courant alternatif	Wechselstrom-Tachogenerator	тахогенератор переменного тока	generatore tachimetrico a corrente alternata	generatriz m tacométrica de corriente alterna	交流タコジェネレータ kōryū-tako-jenerēta	交流测速发电机 jiaoliu cesufadianji
2	absolute stability	stabilité f absolue	absolute Stabilität	абсолютная устойчивость	stabilità assoluta	estabildad f absoluta	絶対安定 zettai-antei	绝对稳定性 juedui wendingxing
3	acceleration	accélération f	Beschleunigung	ускорение	accelerazione	aceleracion f	加速度 kasokudo	加速度 jiasudu
4	access	accès m (grandeur d'entrée f)	Zugriff	доступ, обращение	accesso	acceso m	アクセス akusesu	存取，访问 cunqu, fangwen
5	access time	temps m d'accès	Zugriffszeit	время доступа, время обращения	tempo di accesso	tiempo m de acceso	アクセス時間 akusesu-jikan	存取时间 cunqu shijian
6	accuracy	précision f	Genauigkeit	точность	accuratezza	exactitud f	精度 seido	准确度，精确度 zhunquedu, jingquedu
7	action	action f	Aktion, Wirkung	воздействие, действие	azione	accion f	動作 dōsa	作用 zuoyong
8	active compensation	compensation f active	aktiveKompensation	активная компенсация	compensazione attiva	compensacion * activa	能動的補償 nōdōteki-hoshō	有源补偿 youyuan buchang
9	active element	organe m actif	aktive Schaltung	активный элемент	elemento attivo	elemento m activo	能動要素 nodō-yōso	有源元件，有源环节 youyuan yuanjian, youyuan huanjie
10	active filter	filtre m actif	aktives Filter	активный фильтр	filtro attivo	filtro * activo	能動フィルタ nodō-firuta	有源滤波器 youyuan lüboqi
11	actual value	valeur f réelle	Istwert	фактическое значение	valore effettivo	valor m actual	実際値 jissau-chi	实际值 shijizhi
12	actuate(v)	actionner (v)	steuern, betätigen, antreiben, schalten	воздействовать, возбуждать	attuare	accionar	操作する sōsa-suru	驱动 qudong
13	actuating signal	signal m d'influence	Steuersignal	сигнал воздействия, сигнал возбуждения	segnale di attuazione	senal f activa	操作信号 sōsa-shingō	驱动信号 qudong xinhao

No.	English	French	German	Russian	Italian	Spanish	Japanese	Chinese
14	actuating variable	grandeur f d'influence	Steuergröße	регулирующая величина	variabile di attuazione	variable f activa	操作変数 操作量 sōsa-hensū sōsaryō	驱动变量 qudong bianliang
15	actuator	actionneur m, organe m de réglage final	Stellglied	привод	attuatore	accionador m	アクチュエータ 操作装置 akuchuēta sōsa-sōchi	驱动器. 执行机构 qudongqi, zhixing jigou
16	adaptive algorithm	algorithme m adaptatif	adaptiver Algorithmus	настраивающийся алгоритм, адаптивный алгоритм	algoritmo adattativo	algoritmo m adaptativo	適応アルゴリズム tekiō-arugorizumu	适应算法 shiying suanfa
17	adaptive control	régulation f adaptative	adaptive Regelung	адаптивное (настраиваемое) управление	controllo adattativo	control m adaptativo- control m autoadaptable	適応制御 tekiō-seigyo	适应控制 shiying kongzhi
18	adaptive filter	filtre m adaptatif	adaptives Filter	адаптивный фильтр	filtro adattativo	filtro m adaptativo	適応フィルタ tekiō-firuta	适应滤波器 shiying lüboqi
19	adaptive system	système m adaptatif	adaptives System	адаптивная система	sistema adattativo	sistema m adaptativo	適応系 tekiō-kei	适应系统 shiying xitong
20	adjacency	caractère m adjacent	angrenzen, unmittelbare Umgebung	смежность	adiacenza	adyacencia f	邻近 linjin	
21	adjoint matrix	matrice f adjointe	adjungierte Matrix	присоединенная матрица	matrice aggiunta	matriz f adjunta	余因子行列 yoinshi-gyoretsu	伴随矩阵 bansui juzhen
22	adjust(v)	ajuster (v)	einstellen, abgleichen, nachstellen	настраивать, налаживать	mettere a punto	ajustar	調整する chōsei-suru	调整（动词）tiaozheng (dongci)
23	adjustable	ajustable	einstellbar	настраиваемый	regolabile	ajustable	調整可能な chōsei-kanō-na	可调整的 ketiaozheng de
24	adjuster	ajusteur m	Einstellvorrichtung	орган настройки	regolatore	ajustador m	調整器 chōseiki	调整器 tiaozhengqi
25	adjustment	mise f au point, ajustement m	Einstellung, Abgleichung	наладка, настройка	messa a punto	ajuste m	調整 chōsei	调整 tiaozheng
26	admittance	admittance f	Scheinleitwert, Admittanz	полная проводимость	ammettenza	admitancia f	アドミッタンス adomittansu	导纳 daona

No.	English	French	German	Russian	Italian	Spanish	Japanese	Chinese
27	alarm system	système m d'alarme	Alarmsystem, Alarmeinrichtung	алармирующая система	sistema di allarme	sistema m de alarma	警報系 keihō-kei	报警系统 baojing xitong
28	algorithmic element	circuit m algorithmique	Teil des Algorithmus	алгоритмический элемент	elemento algoritmico	elemento m algorítmico	算法要素 sanpō-yōso	算法环节，算法元件 suanfa huanjie, suanfa yuanjian
29	algorithmic language	langage m algorithmique	algorithmische Sprache	алгоритмический язык	linguaggio algoritmico	lenguaje m algorítmico	アルゴリズム言語 arugorizumu-gengo	算法语言 suanfa yuyan
30	align(v)	aligner (v)	ausrichten, justieren	выравнивать	allineare	alinear	整合する seigō-suru	对准 duizhun
31	all-pass element	déphaseur m pur	Allpaßglied	фазовый элемент	elemento passa-tutto	(elemento) pasa-todo m	全域通過要素 zen-iki-tsūka-yōso	全通元件，全通环节 quantong yuanjian, quantong huanjie
32	all-pass filter	filtre m passe-tout	Allpaßfilter	фазовый фильтр	filtro passa-tutto	filtro * pasa-todo	全域通過フィルタ zen-iki-tsūka-firuta	全通滤波器 quantong lüboqi
33	alphanumeric display	affichage m alphanumérique	alphanumerische Anzeige	алфавитно-цифровой дисплей	visualizzatore alfanumerico	visualización f alfanumérica (unidad de)	英数字表示装置 eisūji-hyōji-sōchi	字符显示 zifu xianshi
34	amplidyne	amplidyne f	Querfeldmaschine, Amplidyne	амплидин, электромашинный усилитель	amplidina	amplidina	アンプリダイン anpuridain	电机放大机 dianji fangdaji
35	amplification	amplification f	Verstärkung	коэффициент усиления, усиление	amplificazione	amplificación f	増幅度 zōfukudo	放大 fangda
36	amplifier	amplificateur m	Verstärker	усилитель	amplificatore	amplificador m	増幅器 zōfukuki	放大器 fangdaqi
37	amplify(v)	amplifier (v)	verstärken	усиливать	amplificare	amplificar	増幅する zōfuku-suru	放大 (动词) fangda (dongci)
38	amplifying element	amplificateur m	Verstärkungseinrichtung	усилитель	elemento amplificatore	elemento m amplificador	増幅要素 zōfuku-yōso	放大环节 fangda huanjie
39	amplitude	amplitude f	Amplitude	амплитуда	ampiezza	amplitud f	振幅 shinpuku	幅值 fuzhi

	English	French	German	Russian	Italian	Spanish	Japanese	Chinese
40	amplitude distortion	distorsion f d'amplitude	Amplitudenverzerrung	амплитудное искажение	distorsione di ampiezza	distorsión f de amplitud	振幅歪 shinpuku-hizumi	幅值畸变 fuzhi jibian
41	amplitude locus	lieu m d'amplitude	Amplitudenortskurve, Amplitudenverlauf	амплитудный годограф	diagramma delle ampiezze	lugar m (geométrico) de amplitudes	振幅軌跡 shinpuku-kiseki	幅值轨迹 fuzhi guiji
42	amplitude response	réponse f en amplitude	Amplitudenantwort, Amplitudengang	амплитудная характеристика	risposta in ampiezza	respuesta f en amplitud	振幅応答 shinpuku-ōtō	幅值响应 fuzhi xiangying
43	analog computer	calculateur m analogique, calculatrice f analogique	Analogrechner	аналоговая вычислительная машина	calcolatore analogico	computador m analógico	アナログ計算機 anarogu-keisanki	模拟计算机 moni jisuanji
44	analog control	régulation f analogique	analoge Regelung (Steuerung)	аналоговое управление (регулирование)	controllo analogico	control m analógico	アナログ制御 anarogu-seigyo	模拟控制 moni kongzhi
45	analog-digital converter	(US)convertisseur m analogique-numérique	Analog-Digitalwandler	аналого-цифровый преобразователь (АЦП)	convertitore analogico-digitale	conversor m analógico-digital-convertidor m analógico-digital	アナログ-ディジタル変換器 anarogu-dijitaru-henkanki	模数转换器 moshu zhuanhuanqi
46	analog signal	signal m analogique	Analogsignal	непрерывный сигнал, аналоговый сигнал	segnale analogico	señal f analógica	アナログ信号 anarogu-shingō	模拟信号 moni xinhao
47	AND element	circuit m ET	UND-Glied (od. Tor)	элемент И	elemento AND	elemento f y	論理積要素 ronriseki-yōsō	"与"元件 "yu" yuanjian
48	AND operation	opération f ET	UND-Operation (od. VerknApfung)	операция И	operazione AND	operación f y	論理演算 ronriseki-enzan	"与"运算, "与"操作 "yu" yunsuan, "yu" caozuo
49	angular acceleration	accélération f angulaire	Winkelbeschleunigung	угловое ускорение	accelerazione angolare	aceleración f angular	角加速度 kaku-kasokudo	角加速度 jiao jiasudu
50	angular deviation	écart m angulaire	Winkelabweichung	угловое рассогласование	deviazione angolare	desviación f angular	(角位置)制御偏差 (kakuichi) seigyo-hansa	角偏差 jiao piancha
51	angular frequency	pulsation f, fréquence f angulaire	Kreisfrequenz	угловая частота	frequenza (pulsazione) angolare	frecuencia * Fourier angular	角周波数 kaku-shūhasū	角频率 jiao pinlü
52	angular measurement	mesure f d'angle	Winkelmessung	угловое измерение	misura angolare	medida/medición angular	角度計測 kakudo-keisoku	角测量 jiao celiang

No.	English	French	German	Russian	Italian	Spanish	Japanese	Chinese
53	angular momentum	moment m cinétique, angulaire	Drehmoment, Drehimpuls	угловой момент	momento angolare	momento * angular	角度モーメンタム kakudo-mōmentamu	角动量 jiao dongliang
54	angular position	position f angulaire	Winkellage	угловое положение	posizione angolare	posición f angular	角位置 kakuichi	角位置 jiao weizhi
55	angular velocity	vitesse f angulaire	Winkelgeschwindigkeit	угловая скорость	velocità angolare	velocidad f angular	角速度 kaku-sokudo	角速度 jiao sudu
56	aperiodic damping	amortissement m sur critique	aperiodische Dämpfung	апериодическое демпфирование	smorzamento aperiodico	amortiguamiento m aperiódico	非周期減衰 hishūki-gensui	非周期阻尼 fei zhougi zuni
57	artificial intelligence (AI)	intelligence f artificielle (I.A.)	künstliche Intelligenz (KI)	искусственный интеллект	intelligenza artificiale (AI)	inteligencia f artificial	人工頭脳 jinkōzunō	人工智能 ren'gong zhineng
58	assembler	programme m d'assemblage	Assembler	ассемблер	assemblatore	ensamblador m	アセンブラ asenbura	汇编程序 huibian chengxu
59	assembly language	langage m d'assemblage	Assemblersprache	язык ассемблера	linguaggio assemblatore	ensamblador m lenguaje m ensamblador	アセンブラ (記号変換ルーチン) asenbura (kigō-henkan-rūchin)	汇编语言 huibian yuyan
60	astatic control	régulation f statique	astatische Regelung	астатическое регулирование	controllo astatico (asintoticamente non polarizzato)	regulación f astática	無定位制御 muteii-seigyo	无静差控制 wujingcha kongzhi
61	astatic controller	régulateur m statique	astatischer Regler (kein P-Anteil)	астатический регулятор	controllore astatico (asintoticamente non polarizzato)	regulador m astático	無定位調節器 muteii-chōsetsuki	无静差控制器 wujingcha kongzhiqi
62	asymptotic approximation	approximation f asymptotique	asymptotische Näherung od. Annäherung	асимптотическая аппроксимация, асимптотическое приближение	approssimazione asintotica	aproximación f asintótica	漸近近似 zenkin-kinji	渐近, 逼近 jianjin, bijun
63	asynchronous sequential logic	logique f séquentielle asynchrone	asynchrone sequentielle (Ablaufsteuerung)	асинхронная последовательная логика	logica sequenziale asincrona	lógica * secuencial asincrona	非同期式逐次論理 hidōkishiki-tsuiji-ronri	异步时序逻辑 yibu shixu luoji
64	attenuate (v)	affaiblir (v)	dämpfen	ослаблять	attenuare	atenuar	減衰する gensui-suru	衰减 (动词) shuaijian (dongci)
65	attenuation	affaiblissement m	Dämpfung	ослабление, затухание	attenuazione	atenuación f	減衰 gensui	衰减 shuaijian

#	English	French	German	Russian	Italian	Spanish	Japanese	Chinese
66	attenuator	dispositif m d'affaiblissement	Dämpfungsglied	ослабитель, аттенюатор	attenuatore	atenuador *	減衰器 gensui-ki	衰减器 shuaijianqi
67	attitude control	commande f, régulation f de l'orientation	Lageregelung	управление положением (относительно земных осей)	controllo di assetto	control de orientación	姿勢制御 shisei-seigyo	姿态控制 zitai kongzhi
68	attitude gyro	gyroscope m d'orientation	Lagenkreisel	гироскоп автопилота	giroscopio	giroscopio m de posición giroscopio m de orientación	姿勢ジャイロ shisei-jairo	姿态陀螺 zitai tuoluo
69	autocorrelation	auto-corrélation f	Autokorrelation	корреляция	autocorrelazione	autocorrelación f	自己相関 jiko-sōkan	自相关 zixiangguan
70	autocorrelation function	fonction f d'auto-corrélation	Autokorrelationsfunktion	корреляционная функция	funzione di autocorrelazione	función f de autocorrelación	自己相関関数 jikosōkan-kansū	自相关函数 zixiangguan hanshu
71	automata theory	théorie f des automates	Automatentheorie	теория автоматов	teoria degli automi	teoria f de autómatas	オートマタ理論 ōtomata-riron	自动机理论 zidongji lilun
72	automate (v)	automatiser (v)	automatisieren	автоматизировать	automatizzare	automatizar	自動化する jidōka-suru	使自动化 (动词) shizidonghua (dongci)
73	automatic	automatique	automatisch	автоматический	automatico	automático/a	自動の 自動的 jidō-no jidō-teki	自动的 zidongde
74	automatic closed-loop control system	système m de réglage automatique	(automatischer) Regelkreis	замкнутая система автоматического управления	sistema di controllo automatico ad anello chiuso	sistema m automático de control en anillo cerrado	自動閉回路制御系 jidō-heikairo-seigyokei	自动闭环控制系统 zidong bihuan kongzhi xitong
75	automatic control (closed loop)	régulation f automatique, régulation f	(automatische) Regelung	автоматическое управление (замкнутое)	controllo automatico (ad anello chiuso)	control m automático (anillo cerrado)	自動制御 (閉回路) jidō-seigyo	自动控制 (闭环) zidong kongzhi (bihuan)
76	automatic control (open loop)	commande f automatique	(automatische) Steuerung	автоматическое управление (разомкнутое)	controllo automatico (ad anello aperto)	control m automático (anillo abierto)	自動制御 (開回路) jidō-seigyo	自动控制 (开环) zidong kongzhi (kaihuan)
77	automatic control engineering	automatique f	Regelungstechnik	автоматическое управление	ingegneria del controllo (automatico)	ingeniería f de control automático	自動制御工学 jidō-seigyo-kōgaku	自动控制工程 zidong kongzhi gongcheng
78	automatic control system	système m de commande automatique	Regelungssystem	система автоматического управления	sistema di controllo automatico	sistema m automático de control	自動制御系 jidō-seigyokei	自动控制系统 zidong kongzhi xitong

No.	English	French	German	Russian	Italian	Spanish	Japanese	Chinese
79	automatic controller	régulateur m	Regler	автоматический контроллер, автоматический регулятор	controllore automatico	controlador m automático	自動調節器 jidō-chōsetsuki	自动控制器 zidong kongzhiqi
80	automatic operation	automatisme m	automatischer Betrieb	автоматическое действие	funzionamento automatico	funcionamiento m automático	自動操作 jidō-sōsa	自动操作，自动运行 zidong caozuo, zidong yunxing
81	automatic process control (closed loop)	commande f automatique d'un processus (en boucle fermée)	Prozeßregelung	автоматическое управление процессом (в замкнутой системе)	controllo automatico dei processi (ad anello chiuso)	control m automático de procesos (anillo cerrado)	自動プロセス制御（閉回路）jidō-purosesu-seigyo	自动过程控制（闭环）zidong guocheng kongzhi (bihuan)
82	automatic process control (open loop)	commande f automatique d'un processus (en boucle ouverte)	Prozeßsteuerung	автоматическое управление процессом (в разомкнутой системе)	controllo automatico dei processi (ad anello aperto)	control m automático de procesos (anillo abierto)	自動プロセス制御（開回路）jidō-purosesu-seigyo	自动过程控制（开环）zidong guocheng kongzhi (kaihuan)
83	automatic regulator	régulateur m automatique	Regler, selbsttätiger Ausgleicher	автоматический регулятор	regolatore automatico	regulador m automático	自動調整器 jidōu-cyōseiki	自动调节器 zidong tiaojieqi
84	automatic restart	redémarrage m automatique	selbsttätiger Wiederanlauf	автоматический рестарт	riavviamento automatico	rearranque automático	自動再始動 jidō-saishidō	自动再启动 zidong zaiqidong
85	automatic sequence control	automatisme m de séquence	selbsttätige Folgesteuerung	программное управление, автоматическое последовательное управление	controllo automatico delle sequenze	control m automático secuencial	自動シーケンス制御 jidō-shikensu-seigyo	自动顺序控制 zidong shunxu kongzhi
86	automatic testing	vérification f automatique	automatisches Testen (od. Prüfen)	автоматическое тестирование	saggiatura automatica	verificación f automática	自動試験 jidō-shiken	自动测试，自动检测 zidong ceshi, zidong jiance
87	automatics	automatismes m.pl	Automatisierung (Allgemeinbegriff)	автоматика тестирование	automatica	automática f	オートメーション	自动学 zidongxue
88	automation	automatisation f	Automatisierung	автоматизация	automazione	automatización f	オートメーション ōto-meishon	自动化 zidonghua
89	automaton	automate m	Automat	автомат	automa	autómata m	オートマトン ōtomaton	自动机 zidongji
90	autotransductor	transducteur m auto-excité	Transduktor in Sparschaltung	авто-трансформаторный магнитный усилитель	autotrasduttore	autotransductor m	単巻トランスダクタ tan-maki-toransudakuta	自馈和电抗器 zibaohe diankangqi
91	availability	disponibilité f	Verfügbarkeit, Betriebsbereitschaft	готовность	disponibilità	disponibilidad f	有効性 yūkōsei	可用性 keyongxing

	English	French	German	Russian	Italian	Spanish	Japanese	Chinese
92	average value	valeur f moyenne	Mittelwert	среднее значение	valor medio	valor m medio	平均値 heikin-chi	平均值 pingjunzhi
93	averaging control	régulation f de moyenne	(Mittelwert-) Regelung, -Steuerung	усреднённое управление (регулирование)	controllo in media	control m de promediación	平均化制御 heikinka-seigyo	平均控制 pingjun kongzhi
94	back-up controller	régulateur m de repli, de secours	Back-up-Regler, Reserveregler	резервный регулятор, резервный контроллер	controllore di riserva	controlador m de reserva	バックアップ制御装置 bakku-appu-seigyo-sōchi	备用控制器 beiyong kongzhiqi
95	back-up system	système m de réserve, de rechange	Reservesystem, Ersatzsystem, Back-up-System	резервная система	sistema di riserva	sistema de salvaguarda (reserva)	バックアップシステム bakku-appu-shisutemu	备用系统 beiyong xitong
96	backlash	jeu m	Lose, toter Gang	люфт	lasco	juego m	バックラッシュ bakkurasshu	间隙 jianxi
97	ball-and-disc integrator	intégrateur m à disque et bille	Kugelscheiben-Integrator	шаровой фрикционный интегратор	integratore a sfera e disco	integrador de bola y disco	ボール・円盤形積分器 bōru-enbangata-sekibunki	球盘积分器 qiupan jifenqi
98	ball resolver	résolveur m à bille	(Kugel-) Drehmelder, Wegmeßgeber	шаровой решающий прибор	resolver a sfera	resolver de bola	球レゾルバ kyū-rezoruba	球形解算器 qiuxing jiesuanqi
99	band-pass filter	filtre m à bande passante, passe-bande	Bandpaßfilter	полосовой фильтр	filtro passa-banda	filtro * pasa-banda	帯域通過フィルタ tai-iki-tsūka-firuta	带通滤波器 daitonglüboqi
100	band-width	largeur f de bande	Bandbreite	ширина полосы	larghezza di banda	ancho m de banda	帯域幅 tai-iki-haba	带宽 daikuan
101	'bang-bang'	plus ou moins	Zweipunkt, Bang-Bang	трёхпозиционное воздействие	"bang-bang"	bang-bang	バンバン ban-ban	棒-棒 bang-bang
102	batch control	régulation f discontinue	Regelung eines Chargenprozesses	циклическое управление, управление циклическими процессами	controllo a lotti	control m por lotes	バッチ制御 bacchi-seigyo	批控制 pikongzhi
103	batch mode	fonctionnement m discontinu	Chargenbetrieb	пакетная работа, пакетный режим	funzionamento a lotti	modalidad f por lotes	バッチモード bacchi-mōdo	批方式 pifangshi
104	bias winding	enroulement m de polarisation	Vormagnetisierungs-wicklung	обмотка смещающего подмагничивания	avvolgimento di polarizzazione	arrollamiento m de polarización	バイアス巻線 baiasu-makisen	偏置绕组 pianzhi raozu

No.	English	French	German	Russian	Italian	Spanish	Japanese	Chinese
105	bilinear system	système m bilinéaire	bilineares System	билинейная система	sistema bilineare	sistema m bilineal	sō-senkei-kei 双線形系	shuangxianxing xitong 双线性系统
106	binary control	régulation f binaire	binäre Steuerung	двоичное управление	controllo binario	control m binario	nishin-seigyo 二進制御	erjinzhi kongzhi 二进制控制
107	binary element	circuit m binaire	Binärelement	двоичный элемент	elemento binario	elemento m binario	nishin(nichi)-yōso 二進(二値)要素	erjinzhi yuanjian, shuangtai yuanjian 二进制元件，双态元件
108	binary-logic system	système m combinatoire	binäres Schaltsystem	двоично-логическая система	sistema a logica binaria	sistema m de lógica binaria	nishin-ronri-kei 二進論理系	erjinzhi luoji xitong 二进制逻辑系统
109	binary signal	signal m binaire	binäres Signal	двоичный сигнал	segnale binario	señal f binaria	nishin-shingō 二進信号	erjinzhi xinhao 二进制信号
110	binary storage element	organe m de mémoire binaire	binäres Speicherelement	двоичный запоминающий элемент	elemento di memoria binario	elemento m binario de memoria	nisin-kioku-yōso 二進記憶要素	erjinzhi cunchu yuanjian 二进制存储元件
111	bio control	régulation f, commande f biologique	biologische Regelung	управление биологическими объектами	biocontrollo	bio control	seitai-seigyo 生体制御	shengwu kongzhi 生物控制
112	biocybernetics	bio-cybernétique f	Biokybernetik	биокибернетика	biocibernetica	biocibernética f	baio-saibanetikusu バイオサイバネティクス	shengwu kongzhi lun 生物控制论
113	bistable multivibrator	multivibrateur m bistable	bistabiler Multivibrator	бистабильный мультивибратор, триггер	multivibratore bistabile	multivibrador m biestable	ni-antei-maruchinaiburēta 2安定マルチバイブレータ	shuangwentai duoxie zhendangqi 双稳态多谐振荡器
114	bistable trigger element	basculeur m bistable	bistabiles Triggerelement	бистабильный триггер	'trigger' (o generatore comandato di segnale d'avvio) bistabile	(elemento) basculador m biestable (elemento) disparador m biestable	sō-antei-toriga-yōso 双安定トリガ要素	shuangwentai chufa yuanjian 双稳态触发元件
115	block diagram	schéma m fonctionnel	Blockschaltbild	блок-схема	schema a blocchi	diagrama m de bloques, esquema m de bloques, esquema m funcional	burokku-senzu ブロック線図	kuangtu 框图
116	Bode diagram	diagramme m de Bode	Bode-Diagramm	диаграмма Боде	diagramma di Bode	diagrama m de bode	bōdo-senzu ボード線図	bodetu 伯德图
117	Boolean algebra	algèbre f de Boole	Boolesche Algebra	булева алгебра	algebra booleana	algebra m de boole	būru-daisu ブール代数	buer daishu 布尔代数

No.	English	French	German	Russian	Italian	Spanish	Japanese	Chinese
118	Boolean function	fonction f booléenne	Boolesche Funktion	булева функция	funzione booleana	función * Booleana	ブール関数 būru-kansu	布尔函数 buer hanshu
119	boundary value problem	problem problème m aux limites	Randwertproblem	краевая задача	problema ai valori di contorno	problema * de condiciones de contorno	境界値問題 kyōkaichi-mondai	边值问题 bianzhi wenti
120	brake	frein m à induction	Bremse	тормоз	freno	freno	ブレーキ 制動 burēki seidō	制动 zhidong
121	breakpoint	point m anguleux	Abschaltpunkt, Knickpunkt (Kurve)	точка разрыва	punto di rottura	punto m de ruptura	折点 setten	转折点、断点 zhuanzhedian, duandian
122	buffer amplifier	amplificateur m tampon	Trennverstärker	буферный усилитель	amplificatore di collegamento	amplificador m separador amplificador m tampón	緩衝増幅器 kanshō-zōfukuki	缓冲放大器 huanchong fangdaqi
123	buffer storage	mémoire f tampon	Pufferspeicher, Zwischenspeicher	буферная память	memoria di collegamento	acumulador m intermedio registro m intermedio	バッファ記憶 緩衝記憶装置 baffa-kioku kanshō-kioku-sōchi	缓冲存储器 huanchong cunchuqi
124	build-up time	temps m d'établissement	Anstiegszeit, Einschwingzeit	характеристическое время, время нарастания	tempo di costruzione	tiempo m característico tiempo m de subida		建立时间 jianli shijian
125	calculator	calculateur m, calculatrice f	Rechner, Rechengerät	калькулятор, счетная машина, вычислительный прибор	calcolatrice	calculadora	計算機 keisan-ki	计算器 jisuanqi
126	calibration	étalonnage m	Kalibrierung	калибровка, калибрование	calibrazione	calibración f calibrado m	校正、較正 kōsei	校准、标定 jiaozhun, biaoding
127	cam	came f	Nocken, Steuerscheibe	кулачок	camma	f.a.c.	カム kamu	凸轮 tulun
128	canonical form	forme f canonique	kanonische Darstellung	каноническая форма	forma canonica	forma f canónica	標準形 正準形 hyōjunkei seijunkei	规范型 guifan xingshi
129	capacitor	condensateur m	Kondensator	конденсатор	condensatore	capacitor m condensador m	容量 yōryō	电容器 dianrongqi
130	capstan amplifier	amplificateur m de couple	Drehmomentverstärker	усилитель вращающего момента	amplificatore ad argano	amplificador capstan		主动轮放大器 zhudonglun fangdaqi

No.	English	French	German	Russian	Italian	Spanish	Japanese	Chinese
131	carrier-frequency amplifier	amplificateur m à courant porteur	Trägerfrequenz-verstärker	усилитель несущей частоты	amplificatore a frequenza portante	amplificador m de frecuencia portadora	搬送周波数増幅器 hansō-shūhasū-zōfukuki	载频放大器 zaipin fangdaqi
132	cascade compensation	correcteur m série	Kaskadenkompensation	каскадная компенсация	compensazione in cascata	compensación f en cascada / compensación f en serie	カスケード補償 kasukēdo-hosyō	串联补偿 chuanlian buchang
133	cascade control	commande f, régulation f en cascade	Kaskadenregelung	каскадное управление (регулирование)	controllo in cascata	control m en cascada	カスケード制御 kasukēdo-seigyo	串级控制 chuanji kongzhi
134	cascade exciter	excitatrice f en cascade	Erregerkaskade (Hauptstromerregung)	каскадный возбудитель	circuito di eccitazione in cascata	excitatriz f en serie	カスケード励磁機 kasukēdo-reijiki	串级激发器，串级励磁器 chuanji jifaqi, chuanji liciqi
135	catastrophe theory	théorie f catastrophe	Katastrophentheorie	теория катастроф	teoria delle catastrofi	teoría f de las catástrofes	破局理論 カタストロフィー理論 hakyoku-riron katasutorōfi-riron	突变论 tubianlun
136	cathode follower	cathodyne f	Kathodenfolger	катодный повторитель	inseguitore catodico	seguidor m catódico	カソードホロア kasōdo-horōa	阴极跟随器 yinji gensuiqi
137	causality	causalité f	Kausalität	каузальность, причинность	causalità	causalidad f	因果性 因果律 ingakankei ingaritsu	因果性 yinguo guanxi, yinguolü
138	cellular automation	automatisation f cellulaire	zellularer Automat	сотовая автоматизация	automazione cellulare	automatización celular		单元自动化 danyuan zidonghua
139	center	centre m	Zentrum	центр	centro	centro m	中心点 chūshinten	中心 zhongxin
140	central processing unit (CPU)	unité f centrale de traitement (U.C.T.)	Zentrale Recheneinheit (CPU)	центральный процессор (ЦП)	unità centrale di elaborazione (CPU)	unidad f central de proceso	中央処理装置 chūō-shori-sōchi	中央处理机 zhongyang chuliji
141	central processor	processeur m central	Zentralrechner	центральный процессор	elaboratore centrale	procesador m central	中央処理装置 chūō-shori-sōchi	中央处理机 zhongyang chuiji
142	centralized control	régulation f centralisée	zentrale Regelung (Steuerung)	централизованное управление	controllo centralizzato	control m centralizado	集中制御 shūchūka-seigyo	集中控制 jizhong kongzhi
143	centrifugal governor	régulateur m à force centrifuge	Fliehpendelregler	центробежный регулятор	regolatore a forza centrifuga	regulador m centrífugo	遠心調速機 enshin-chōsokuki	离心式调速器 lixinshi tiaosuqi

	English	French	German	Russian	Italian	Spanish	Japanese	Chinese
144	chain	chaîne f	Kette	цепь	catena	cadena f	チェン chen	链，链式 lian, lianshi
145	changeover contacts	contacts m de basculement	Umschaltkontakte	переключающие контакты	contatti di commutazione	contactos m inversores	切換接点 kirikae-setten	转换触点 zhuanhuan chudian
146	changeover gate	commutateur m, porte f d'aiguillage	Umschalttor	переключающий вентиль	porta di commutazione	puerta f inversora	切換ネゲート kirikae-gēto	转换门 zhuanhuan men
147	channel	canal m	Kanal	канал	canale	canal m	チャネル chaneru	通道，信道 tongdao xindao
148	character recognition	reconnaissance f de caractère	Zeichenerkennung	распознавание символов	riconoscimento dei caratteri	reconocimiento m de caracteres	文字認識 moji-ninsiki	字符识别 zifu shibie
149	characteristic curve	courbe f caractéristique, caractéristique f	Kennlinie, charakteristische Kurve	характеристика	curva caratteristica	característica f curva f característica	特性曲線 tokusei-kyokusen	特性曲线 texing quxian
150	characteristic equation	équation f caractéristique	charakteristische Gleichung	характеристическое уравнение	equazione caratteristica	ecuación f característica	特性方程式 tokusei-hōteishiki	特征方程 tezheng fangcheng
151	characteristic polynomium	polynôme m caractéristique	charakteristische Polynom	характеристический полином	polinomio caratteristico	polinomio m característico	特性多項式 tokusei-takōshiki	特征多项式 tezheng duoxiangshi
152	characteristic roots	racines f,pl caractéristiques	charakteristische Wurzeln	характеристический корень	radici caratteristiche	Raíces características	特性根 tokusei-kon	特征根 tezheng gen
153	characteristic time	temps m caractéristique	Zeitkenngröße, Sollzeit	характеристическое время, время нарастания	tempo caratteristico	tiempo m característico	特性時間 tokusei-jikan	特性时间 texing shijian, jianli shijian
154	characteristic vector	vecteur m caractéristique	Eigenvektor	характеристический вектор	vettore caratteristico	vector m característico	特性ベクトル tokusei-bekutoru	特征矢量 tezheng shiliang
155	check valve	vanne f de retenue	Rückschlagventil	стопорный клапан, обратный клапан, невозвратный клапан	valvola d'interdizione	válvula f de retención	チェック弁 chekku-ben	检验阀 jianyanfa
156	choke	bobine f d'arrêt	Drossel, Luftklappe	дроссель	strozzatura	estrangulador	チョーク chōku	扼流圈 eliuquan

No.	English	French	German	Russian	Italian	Spanish	Japanese	Chinese
157	chopper	hacheur m	Zerhacker, Unterbrecher	прерыватель	'chopper', interruttore	rebanador m, vibrador m	チョッパ choppa	斩波器，振动换流器 zhanboqi, zhendong huanliuqi
158	chopper amplifier	amplificateur m à découpages	Zerhackerverstärker	вибраторный усилитель, усилитель с виброрпреобразователем	amplificatore a modulazione	amplificador chopper	チョッパ増幅器 choppa-zōfukuki	斩波放大器 zhanbo fangdaqi
159	circuit	circuit m	Stromkreis, Schaltkreis	цепь	circuito	circuito m	回路 kairo	电路 dianlu
160	closed loop	boucle f fermée	geschlossener Kreis	замкнутый контур	anello chiuso	anillo m cerrado, bucle m cerrado, cadena f cerrada	閉回路 tojita-rūpu	闭环 bihuan
161	closed-loop control system	système m de régulation en boucle fermée	Regelkreis	замкнутая система управления	sistema di controllo ad anello chiuso	sistema m de control en anillo cerrado	閉回路制御系 heikairo-seigyokei	闭环控制系统 bihuan kongzhi xitong
162	closed-loop gain	gain m en boucle fermée	Regelkreisverstärkung	коэффициент усиления при замкнутом контуре	guadagno ad anello chiuso	ganancia f en anillo cerrado	閉回路ゲイン heikairo-gein	闭环增益 bihuan zengyi
163	closed-loop phase angle	déphasage m en boucle fermée	Phasenwinkel des geschlossenen Systems	фазовый сдвиг при замкнутом контуре	angolo di fase ad anello chiuso	desfasaje m en anillo cerrado	閉回路位相角 heikairo-isōkaku	闭环相角 bihuan xiangjiao
164	closed-loop system	système m en boucle fermée	geschlossenes System, Regelkreissystem	замкнутая система	sistema ad anello chiuso	sistema m en anillo cerrado	閉回路系 heirūpu-kei	闭环系统 bihuan xitong
165	closed-loop transfer function	transmittance f en boucle fermée	Übertragungsfunktion des geschlossenen Kreises	передаточная функция замкнутой системы	funzione di trasferimento ad anello chiuso	función f de transferencia en anillo cerrado	閉回路伝達関数 heikairo-dentatsu-kansū	闭环传递函数 bihuan chuandihanshu
166	clutch	embrayage m	Kupplung	сцепление	innesto	embrague	クラッチ kuratti	离合器 liheqi
167	coarse-fine control	régulation f, approche f, précision f	Grob-Fein-Steuerung (Regelung)	"грубо-точное" управление, двухступенчатое управление	controllo grossolano-fine	control m grueso-fino	精粗制御 seiso-seigyo	粗一精控制 cu-jing kongzhi
168	coarse-fine relay	commutateur m, approche f, précision f	Grob-Fein-Relais (Fernsteuerung)	реле переключения точности	'relay' (di commutazione) grossolano-fine	relé m grueso-fino	精粗リレー seiso-rirē	粗一精继电器 cu-jing jidianqi
169	coarse-fine switch	commutateur m, approche f, précision f	Grob-Fein-Schalter (Fernsteuerung)	переключатель точности	commutatore grossolano-fine	conmutador m grueso-fino	精粗スイッチ seiso-swicchi	粗一精开关 cu-jing kaiguan

No.	English	Français	Deutsch	Русский	Italiano	Español	日本語	中文
170	code	code *m*	Code, Schlüssel	код	codice	código *m*	符号 fugō	码, 代码 ma, daima
171	code converter	transcodeur *m*	Codeumsetzer	преобразователь кодов	convertitore di codice	convertidor *m* de código	符号变换器 fugō-henkanki	代码转换器, 译码器 daima zhuanhuanqi, yimaqi
172	coder	codeur *m*	Codierer, Verschlüssler	кодирующее устройство	codificatore	codificador *m*	コーダ kōda	编码器, 编码员 bianmaqi, bianmayuan
173	coefficient of stability	coefficient *m* de stabilité	Stabilitätskoeffizient	коэффициент стабильности	coefficiente di stabilità	coeficiente *m* de estabilidad	安定系数 antei-keisū	稳定系数 wending xishu
174	cognitive system	système *m* cognitif	kognitives System	познавательная система	sistema cognitivo	sistema cognitivo	認識系 ninshiki-kei	识别系统 shibie xitong
175	coherency	cohérence *f*	Kohärenz	когерентность	coerenza	coherencia *f*	コヒーレンシー kohīrensi	关联性 guanlian xing
176	colored noise	bruit *m* coloré	farbiges Rauschen	окрашенный шум	rumore colorato	ruido *m* coloreado	有色雑音 yūshoku-zatsuon	有色噪声 youse zaosheng
177	combinational circuit	circuit *m* combinatoire	Schaltnetz	комбинационная цепь	circuito combinatorio	circuito *m* combinatorio	複合回路 fukugō-kairo	组合电路 zuhe dianlu
178	combinatorial logic	logique *f* combinatoire	Schaltlogik	комбинаторная логика	logica combinatoria	lógica combinatoria	組み合わせ論理 kumiawase-ronri	组合逻辑 zuhe luoji
179	command	signal *m* de commande	(Steuerungs- od. Schalt-)Befehl	команда	comando	orden *f*	目標値 mokuhyō-chi	指令 zhiling
180	command control	contrôle-commande *m*	Steuerung, Schaltung	управление войсками, командование	controllo a comando	control	目標値制御 mokuhyō-chi-seigyo	指令控制 zhiling kongzhi
181	command signal	signal *m* de commande	Steuersignal, Schaltsignal	команда	segnale di comando	señal *f* de mando	目標信号 mokuhyō-chi-singō	指令信号 zhiling xinhao
182	command variable	grandeur *f* de commande	Steuervariable, Schaltvariable	управляющее воздействие	variabile di comando	variable *f* de mando	目標値 mokuhyō-chi	指令变量 zhiling bianliang

No.	English	French	German	Russian	Italian	Spanish	Japanese	Chinese
183	communication system	réseau m de communication	Nachrichtensystem, Kommunikationssystem	система связи, система коммуникации	sistema di comunicazione	sistema m de comunicación	通信系 tsūshin-kei	通信系统 tongxin xitong
184	companion matrix	matrice f compagne	Begleitmatrix (Frobeniusform)	сопровождающая матрица	matrice compagna	matriz f asociada	同伴行列 dōhan-gyōretsu	伴随矩阵 bansui juzhen
185	comparator	comparateur m	Vergleicher	элемент сравнения	comparatore	comparador m	比较器 hikaku-ki	比较器, 比较电路 bijiaoqi, bijiao dianlu
186	comparing element	comparateur m	Vergleichsglied	элемент сравнения	elemento di comparazione	elemento m de comparación	比較要素 hikaku-yōso	比较环节, 比较元件 bijiao huanjie, bijiao yuanjian
187	compatibility	compatibilité f	Kompatibilität,	совместимость	compatibilità	compatibilidad f	両立性 適合性 ryōritsusei tekigōsei	相容性, 兼容性 xiangrongxing, jianrongxing
188	compatible	compatible	kompatibel, verträglich	совместимый	compatibile	compatible	同等 dōtō	兼容的 jianrongde
189	compensating element	circuit m de compensation	Kompensationsglied	компенсирующий элемент	elemento compensatore	(elemento) compensador m	補償要素 hoshō-yōso	补偿环节 buchang huanjie
190	compensating feedback	réaction f compensatrice	Rückführkompensation	компенсирующая обратная связь	compensatore in retroazione	realimentación f compensadora	補償用フィードバック hoshō-yō-fīdobakku	补偿反馈 buchang fankui
191	compensating feedforward	action f compensatrice	Serienkompensation (vorwärts)	компенсирующая прямая связь	compensatore ad azione diretta	anticipación f compensadora	補償用フィードフォワード hoshō-yō-fīdofowādo	补偿前馈 buchang qiankui
192	compensating winding	enroulement m de compensation	Kompensationswicklung	компенсационная обмотка	avvolgimento di compensazione	arrollamiento de compensación	補償巻線 hoshō-makisen	补偿绕组 buchang raozu
193	compensation	compensation f	Kompensation, Ausgleich	компенсация	compensazione	compensación f	補償 hoshō	补偿 buchang
194	compiler	compilateur m	Kompilierer, Übersetzer (Programm)	компилятор	compilatore	compilador m	コンパイラ konpaira	编译程序 bianyi chengxu
195	complement	complément m	Komplement, Ergänzung	дополнение	complemento	complemento m	補数 hosū	补码 buma

No.	English	français	Deutsch	русский	italiano	español	日本語	中文
196	complementary feedback	retour m complémentaire	zusätzliche Rückführung	дополнительная обратная связь	retroazione complementare	realimentación f complementaria	補助帰還 hojo-kikan	辅助反馈 fuzhu fankui
197	complementary function	fonction f complémentaire	Komplementärfunktion, Zusatzfunktion	дополнительная функция	funzione complementare	función f complementaria	補関数 hokansū	余函数 yu hanshu
198	complete controllability	gouvernabilité f entière	vollständige Steuerbarkeit	полная управляемость	completa controllabilità	controlabilidad f total	完全可制御性 kanzen-ka-seigyo-sei	完全可控性 wanquan kekongxing
199	complex plane	plan m complexe	komplexe Ebene (Gaußsche Zahlenebene)	комплексная плоскость	piano complesso	plano m complejo	複素平面 fukuso-heimen	复平面 fupingmian
200	complex variable	variable f complexe	komplexe Variable	комплексная переменная	variabile complessa	variable f compleja	複素変数 fukusosū	复变量 fubianliang
201	compliance	souplesse f	Nachgiebigkeit, Durchbiegung	податливость, согласие	subordinazione	acomodación f	コンプライアンス konpuraiansu	柔顺，顺应 roushun, shunying
202	component	composant m	Bestandteil, Bauelement, Bauteil	элемент, компонент	componente	componente m	構成要素 kōsei-yōso	部件，组元 bujian, zuyuan
203	composite action	action f composite	zusammengesetztes Verhalten	составное действие, сложное действие	azione composita	acción f combinada	複合動作 fukugō-dōsa	组合作用 zuhe zuoyong
204	compound action	action f composée	Verbundverhalten	комбинированное воздействие	azione combinata	acción f compuesta	複合動作 fukugō-dōsa	复合作用 fuhe zuoyong
205	compound controller	régulateur m à actions composées	(Verbund-) Regler	составной регулятор, сложный регулятор	controllore ad azione combinata	controlador m compuesto	複合制御装置 fukugō-seigyo-sōchi	复合控制器 fuhe kongzhiqi
206	compounding feedback	réaction f composite	zusammengesetzte Rückführung	многокомпонентная обратная связь	retroazione combinata	realimentación f compuesta	複合フィードバック fukugō-fīdobakku	复合反馈 fuhe fankui
207	compounding feedforward	action f directe composite	zusammengesetzter Vorwärtszweig	многокомпонентная прямая связь	azione combinata diretta	anticipación f compuesta	複合フィードフォワード fukudo-fīdo-fowādo	复合前馈 fuhe qiankui
208	computability	calculabilité f	Berechenbarkeit	вычислимость, исчисляемость	computabilità	calculabilidad f computabilidad f	計算可能性 keisan-kanōsei	可计算性 kejisuanxing

No.	English	French	German	Russian	Italian	Spanish	Japanese	Chinese
209	computational algorithm	algorithme m de calcul	Rechenalgorithmus	вычислительный алгоритм	algoritmo computazionale	algoritmo * computacional	計算アルゴリズム keisan-arugorizmu	计算算法 jisuan suanfa
210	compute (v)	calculer (v)	rechnen, berechnen	вычислять	calcolare	calcular computar	計算する keisan-suru	计算 jisuan
211	computer	calculateur m, calculatrice f, ordinateur m	Rechenmaschine, Computer	вычислительная машина	calcolatore	calculador m computador/a m/f ordenador m	計算機 keisanki	计算机 jisuanji
212	computer-aided design (CAD)	conception f assistée par ordinateur (C.A.O.)	rechnergestützter Entwurf (CAD)	автоматизированное проектирование	progettazione assistita da calcolatore (CAD)	diseño m asistido por computador (dac)	計算機利用設計 keisanki-enyō-sekkei	计算机辅助设计 jisuanji fuzhu sheji
213	computer-aided engineering (CAE)	ingénierie f assistée par ordinateur (I.A.O.)	rechnergestütztes Engineering (CAE)	компьютеризованное конструирование	ingegnerizzazione assistita da calcolatore (CAE)	ingeniería asistida por computador	計算機利用工学 keisanki-enyō-kōgaku	计算机辅助工程 jisuanji fuzhu gongcheng
214	computer-aided manufacturing (CAM)	fabrication f, production f assistée par ordinateur (F.A.O), (P.A.O.)	rechnergestütze Fertigungssteuerung (CAM)	автоматизированное производство, компьютеризованное производство	produzione assistita da calcolatore (CAM)	fabricación f asistida por computador (fac)	計算機利用生産 keisanki-enyō-seisan	计算机辅助制造 jisuanji fuzhu zhizao
215	computer-aided test (CAT)	contrôle m, vérification f assisté(e) par ordinateur (V.A.O.)	rechnergestütztes Prüfen (CAT)	компьютеризованное испытание, компьютеризованная проверка	prova assistita da calcolatore (CAT)	verificacion asistida por computador	計算機利用試験 keisanki-enyō-shiken	计算机辅助测试 jisuanji fuzhu ceshi
216	computer control	régulation f par calculateur	rechnergeführte Regelung	компьютерное управление	controllo mediante calcolatore	control m por computador	計算機制御 keisanki-seigyo	计算机控制 jisuanji kongzhi
217	computer-controlled system	système m réglé par calculateur	rechnergeführtes System	система с компьютерным управлением	sistema controllato mediante calcolatore	sistema m controlado por computador sistema m de control con computador	計算機控制系 keisanki-seigyo-kei	计算机控制系统 jisuanji kongzhi xitong
218	computer graphics	infographie f	Computergraphik, graphische Datenverarbeitung	машинная графика	grafica mediante calcolatore	gráficos por computador	コンピュータグラフィックス konpyūta-gurafikkusu	计算机图示学 jisuanji tushixue
219	computer hardware	matériel m d'ordinateur	Rechner-Hardware	вычислительное оборудование	hardware del calcolatore	material (hardware) del computador	計算機ハードウェア keisanki-hādowuea	计算机硬件 jisuanji yingjian
220	computer program	programme m d'ordinateur	Rechnerprogramm	программа (вычислительной машины)	programma del calcolatore	programa de computador	計算機プログラム keisanki-puroguramu	计算机程序 jisuanji chengxu
221	computer software	logiciel m d'ordinateur	Rechner-Software	программное обеспечение	software del calcolatore	logical	計算機ソフトウェア keisanki-sofutowuea	计算机软件 jisuanji ruanjian

No.	English	French	German	Russian	Italian	Spanish	Japanese	Chinese
222	computer subroutine	sous-routine f	Rechner-Subroutine	подпрограмма	sottoprogramma del calcolatore	subprograma de computador	計算機サブルーチン keisanki-sabūruchin	计算机子程序 jisuanji zichengxu
223	computing element	organe m de calcul	Rechnereinheit	вычислительный элемент	elemento di calcolo	elemento m de cálculo	演算要素 enzan-yōso	计算环节, 计算元件 jisuan huanjie, jisuan yuanjian
224	computing linkage	mécanisme m de calcul	Rechnerverbindung	вычислительная связь	connessione di calcolo	montaje	演算リンク機構 enzan-rinku-kikō	计算连接 jisuan lianjie
225	concentrator	concentrateur m	(Daten-)Konzentrator	концентратор	concentratore	concentrador m		集中器 jizhongqi
226	conditional stability	stabilité f conditionnelle	bedingte Stabilität	условная устойчивость	stabilità condizionata	estabilidad f condicional	条件付安定性 jōkentsuki-anteisei	条件稳定性 tiaojian wendingxing
227	configuration	configuration f	Konfiguration (gerätetechnische Anordnung)	конфигурация	configurazione	configuración f	配置 構成 haichi kōsei	组态, 构形 zutai, gouxing
228	configuration control	régulation f de configuration	Konfigurationsüberwachung (Programmodule)	конфигурационное управление	controllo di configurazione	control m de configuración		组态控制, 构形控制 zutai kongzhi, gouxing kongzhi
229	configuration stability	stabilité f de configuration	Stabilität der Konfiguration (Programmodule)	конфигурационная стабильность	stabilità di configurazione	estabilidad de la configuración		组态稳定性, 构形稳定性 zutai wendingxing, gouxing wendingxing
230	conformal mapping	correspondance f conforme	konforme Abbildung	конформное отображение	trasformazione conforme	representación f conforme	等角写像 tōkaku-shazō	保角映射 baojiao yingshe
231	conjugate roots	racine f conjuguée	konjugiert komplexe Wurzeln	сопряженные корни	radici coniugate	raices f conjugadas	共役根 kyōyaku-kon	共轭根 gongegen
232	conjunction	conjonction f	Verbindung, Konjunktion	конъюнкция	congiunzione	conjunción f	論理積 gōsetsu	"与", 合取 "yu", hequ
233	connective instability	instabilité f associative	Verbundinstabilität	связующая нестабильность	instabilità connettiva	inestabilidad conectiva f		连结不稳定性 lianjie buwendingxing
234	connective stability	stabilité f associative	Verbundstabilität	связующая стабильность	stabilità connettiva	estabilidad conectiva		连结稳定性 lianjie wendingxing

#	English	French	German	Russian	Italian	Spanish	Japanese	Chinese
235	console	pupitre m	Schaltpult, Steuerpult	консоль, пульт	consolle	cónsola f	操作卓 端末装置 sōsataku tanmatsu-sōchi	控制台 kongzhitai
236	constant of inertia	constante f d'inertie	Trägheitskonstante	постоянная инерции	costante d'inerzia	constante f de inercia	慣性定数 kansei-jōsū	惯性常数 guanxing xishu
237	constrained parameter	paramètre m limité	beschränkte Parameter	параметр с ограничениями	parametro vincolato	parámetro m forzado	拘束パラメータ kōsoku-paramēta	约束参量 yueshu canliang
238	constraint	limitation f	Beschränkung, Nebenbedingung, Zwangsbedingung	ограничение	vincolo	ligadura f restricción f	拘束 kōsoku	约束 yueshu
239	contact	contact m	Kontakt, Berührung	контакт	contatto	contacto m	接点 setten	触点、接触 chudian, jiechu
240	contact pick-off	capteur m à contact	Kontaktabnehmer	контактный датчик	captatore a contatto	selección por contacto	接点 setten	接触拾出端 jiechu jianchuduan
241	content addressable storage	stockage m en mémoire adressable	Asoziativspeicher, inhaltsadressierbarer Speicher	ассоциативная память	memoria a contenuto indirizzabile	almacenamiento direccionable por el contenido	連想記憶器 rensō kioku	相联存储器 xianglian cunchuqi
242	continuous action	action f permanente, continue	kontinuierliche (Signal-)Wirkung	непрерывное воздействие	azione continua	acción f continua	連続動作 renzoku dōsa	连续作用 lianxu zuoyong
243	continuous-action controller	régulateur m à action continue	kontinuierlich arbeitender Regler	регулятор непрерывного действия	controllore ad azione continua	controlador m de acción continua	連続動作調節器 renzokudōsa-chōsetsuki	连续作用控制器 lianxu zuoyong kongzhiqi
244	continuous control	commande f continue	(zeit-)kontinuierliche Regelung	непрерывное управление (регулирование)	controllo continuo	control m continuo	連続制御 renzoku-seigyo	连续控制 lianxu kongzhi
245	continuous system	système m continu	kontinuierliches System	непрерывная система	sistema continuo	sistema m continuo	連続系 renzoku-kei	连续系统 lianxu xitong
246	continuous variable	grandeur f continue	kontinuierliche (od. Veränderliche) Variable	непрерывная переменная	variabile continua	variable f continua	連続変数 renzoku-hensū	连续变量 lianxu bianliang
247	control (closed loop)	régulation f (boucle fermée), asservissement m	Regelung (geschlossener Kreis)	регулирование (замкнутое), управление (замкнутое)	controllo (ad anello chiuso)	control m (anillo cerrado)	制御 (閉回路) seigyo	控制 (闭环) kongzhi (bihuan)

No.	English	French	German	Russian	Italian	Spanish	Japanese	Chinese
248	control (general term)	commande f, conduite f	Regelung, Steuerung (Sammelbegriff)	регулирование, управление	controllo (termine generale)	control m (término genérico)	制御 (一般用語) seigyo	控制 (一般词) kongzhi (yibanci)
249	control (open loop)	commande f directe	Steuerung (offener Kreis)	регулирование (разомкнутое), управление (разомкнутое)	controllo (ad anello aperto)	control m (anillo abierto)	制御 (開回路) seigyo	控制 (开系) kongzhi (kaihuan)
250	control (v) (closed loop)	régler (v)	regeln (geschlossener Kreis)	регулировать (в замкнутой системе), управлять (в замкнутой системе)	controllare (ad anello chiuso)	controlar (anillo cerrado)	制御する (閉回路) seigyo-suru	控制 (动词) (闭系) kongzhi (dongci) (bihuan)
251	control (v) (general term)	conduire (v), régir (v)	regeln, steuern (Sammelbegriff)	регулировать, управлять	controllare (termine generale)	controlar (término genérico)	制御する (一般用語) seigyo-suru	控制 (动词) (一般词) kongzhi (dongci)
252	control (v) (open loop)	commander (v)	steuern (offener Kreis)	регулировать (в разомкнутой системе), управлять (в разомкнутой системе)	controllare (ad anello aperto)	controlar (anillo abierto)	制御する (開回路) seigyo-suru	控制 (动词) (开系) kongzhi (dongci) (kaihuan)
253	control accuracy	précision f de réglage	Steuerungs - bzw. Regelungsgenauigkeit	точность управления	accuratezza del controllo	exactitud f de control	制御精度 seigyo-seido	控制精度 kongzhi jingquedu
254	control action	action f de réglage	Regel - bzw. Steuerwirkung	управляющее воздействие	azione di controllo	acción f de control	制御動作 seigyo-dōsa	控制作用 konzhi zuoyong
255	control algorithm	algorithme m de réglage	Regelalgorithmus	алгоритм управления	algoritmo di controllo	algoritmo m de control	制御アルゴリズム seigyo-arugorizumu	控制算法 kongzhi suanfa
256	control circuit (closed loop)	circuit m de réglage	Regelkreis (-Schaltung)	цепь управления (замкнутая)	circuito di controllo (ad anello chiuso)	circuito m de control (anillo cerrado)	制御回路 (閉回路) seigyo-kairo (hei-kairo)	控制线路 (闭系) kongzhi xianlu (bihuan)
257	control circuit (open loop)	circuit m de commande	Steuerschaltung	цепь управления (разомкнутая)	circuito di controllo (ad anello aperto)	circuito m de control (anillo abierto)	制御回路 (開回路) seigyo-kairo (kai-kairo)	控制线路 (开系) kongzhi xianlu (kaihuan)
258	control engineering	automatique f	Regelungstechnik	автоматическое управление	ingegneria del controllo	ingeniería f de control	制御工学 seigyo-kōgaku	控制工程 kongzhi gongcheng
259	control equipment	équipement m de commande	Regeleinrichtung	управляющее устройство	apparecchiatura di controllo	equipo m de control	制御装置 seigyo-sōchi	控制装置 kongzhi zhuangzhi
260	control error	erreur f de réglage	Regelfehler	погрешность управления (регулирования)	errore di controllo	error m de control	制御偏差 seigyo-hensa	控制误差 kongzhi wucha

	English	French	German	Russian	Italian	Spanish	Japanese	Chinese
261	control function	fonction f de commande	Regelaufgabe (od. Funktion)	критерий управления, функция управления	funzione di controllo	función de control	制御関数 seigyo-kansū	控制功能. 控制函数 kongzhi gongneng, kongzhi hanshu
262	control instant	instant m de réglage	Momentanwert d. Regelung bzw. Steuerung	момент управления	istante di controllo	instante m de control momento m de control		控制时刻 kongzhi shike
263	control law	loi f de commande	Regelgesetz	алгоритм управления, закон управления	legge di controllo	ley f de control	制御律 seigyosoku	控制律 kongzhilü
264	control loop	boucle f de commande; boucle f d'asservissement	Regelkreis	контур управления	anello di controllo	anillo m de control	制御回路 制御ループ seigyo-kairo seigyo-rūpu	控制回路 kongzhi huilu
265	control panel	panneau m de réglage	Bediengerät, Bedienpult	пульт управления	pannello di controllo	panel m de control	制御パネル seigyo-paneru	控制台 kongzhi pan, kongzhi tai
266	control point	lieu m de réglage, emplacement m de réglage	Sollwert, Kontrollpunkt	точка управления	punto di controllo	punto m de control	制御点 seigyoten	控制点 kongzhi dian
267	control precision	précision f de réglage	Regelgenauigkeit	точность управления (или регулирования)	precisione del controllo	precisión f de control	制御精度 seigyo-seido	控制精度 kongzhi jingdu
268	control range	étendue f de régulation	Regelungsbereich *	диапазон регулирования	gamma (della variabile) di controllo	dominio de regulacion margen m de control	制御範囲 seigyo-han-i	控制范围 kongzhi fanwei
269			Regelbereich *		gamma (della variabile) di controllo	dominio de regulacion margen m de control		控制范围 kongzhi fanwei
270	control station	poste m de commande	Steuer - od. Regelstation	пункт управления	stazione di controllo	puesto m de mando	制御ステーション seigyo-sutēshon	控制站 kongzhizhan
271	control synchro	synchro-machine f de commande	Steuermelder	сельсин-датчик	synchro di controllo	sincro m de control	制御シンクロ seigyo-shinkuro	控制式自整角机 kongzhishi zizhengjiaoji
272	control system (closed loop)	système m de régulation, système m asservi	Regelungssystem, Regelsystem	система управления (замкнутая)	sistema di controllo (ad anello chiuso)	sistema m de control (anillo cerrado)	制御系 (閉回路) seigyokei (hei-kairo)	控制系统 (闭环) kongzhi xitong (bihuan)
273	control system (open loop)	système m de commande directe	Steuerkette	система управления (разомкнутая)	sistema di controllo (ad anello aperto)	sistema m de control (anillo abierto)	制御系 (開回路) seigyokei (kai-kairo)	控制系统 (开环) kongzhi xitong (kaihuan)

	English	French	German	Russian	Italian	Spanish	Japanese	Chinese
274	control unit	élément m de réglage	Regel- od. Steuereinheit	устройство (=блок) управления	unità di controllo	unidad f de control	制御要素 制御ユニット seigyo-yōso seigyo-yunitto	控制单元 kongzhi danyuan
275	control valve	vanne f (de régulation)	Stellventil	клапан управления	valvola di controllo	válvula f de control válvula f de mando	制御弁 seigyo-ben	控制阀 kongzhifa
276	control winding	enroulement m de commande	Steuerwicklung	обмотка управления	avvolgimento di controllo	arrollamiento m de control	制御巻線 seigyo-makisen	控制绕组 kongzhi raozu
277	controllability	gouvernabilité f, commandabilité f	Steuerbarkeit	управляемость	controllabilità	controlabilidad f	可制御性 ka-seigyo-sei	可控性 kekongxing
278	controllability index	indice m de gouvernabilité, de commandabilité	Steuerbarkeitsindex	индекс управляемости	indice di controllabilità	índice m de controlabilidad	可制御性指数 ka-seigyo-shisū	可控性指数 kekongxing zhishu
279	controllable	réglable	steuerbar	управляемый	controllabile	controlable	可制御な kaseigyo-na	可控的 kekongde
280	controlled condition	grandeur f réglée	Regelgröße	управляемое условие	condizione controllata	condición controlada	制御状態 seigyo-jōtai	被控条件，控制条件 beikong tiaojian, kongzhi tiaojian
281	controlled device	organe m de commande	geregelte od. gesteuerte Einrichtung	управляемое устройство, управляемый объект	dispositivo controllato	dispositivo m controlado	被制御装置 hi-seigyo-sōchi	被控装置 beikong zhuangzhi
282	controlled member	organe m commandé	Regel - od. Steuerobjekt	управляемое устройство, управляемый объект	elemento controllato	unidad f controlada	被制御物体 hi-seigyo-buttai	控制对象，受控对象 kongzhi duixiang
283	controlled network	réseau m réglé	geregeltes Netzwerk	управляемая сеть	rete controllata	red controlada	制御回路 seigyo-kairo	被控网络 beikong wangluo
284	controlled system	système m commandé	geregeltes System (Regelstrecke)	управляемая система	sistema controllato	sistema m controlado	制御対象 seigyo-taishō	被控系统 beikong xitong
285	controlled variable	grandeur f réglée	Regelgröße	регулируемая переменная, управляемая переменная	variabile controllata	variable controlada	制御量 seigyo-ryō	被控变量 beikong bianliang
286	controller	régulateur m	Regler	контроллер, регулятор	controllore	controlador m	調節器 コントローラ chōsetsuki kontorōra	控制器 kongzhiqi

	English	French	German	Russian	Italian	Spanish	Japanese	Chinese
287	controlling element	organe *m* de réglage	Stellglied, Stellorgan	управляющий элемент	elemento controllante	elemento *m* de control	調節要素 chōsetsu-yōso	减控环节，减控元件 shikong huanjie, shikong yuanjian
288	controlling machine	machine *f* réglante	regelnde od. steuernde Maschine	управляющая машина	macchina controllante	máquina de control	制御機械 seigyo-kikai	控制机，控制装置 kongzhiji, kongzhi zhuangzhi
289	controlling power station	poste *m* de conduite en puissance	Regelkraftwerk	управляющая электростанция	centrale controllante	planta de potencia de control	制御発電所 seigyo-hatsudensho	控制电站 kongzhi donglizhan
290	controlling system	équipement *m* de commande	Regel- od. Steuereinrichtung	управляющая система	sistema controllante	sistema *m* controlador	制御装置 seigyo-sōchi	减控系统，控制系统 shikong xitong, kongzhi xitong
291	convergent control	système *m* réglant	konvergierende Regelung	комбинированное управление, сходящееся управление	controllo convergente	control *m* convergente	集束制御 syūsoku-seigyo	收敛控制 shoulian kongzhi
292	convergent series	série *f* convergente	konvergente Reihe (math.)	сходящийся ряд	serie convergente	serie * convergente	収束級数 syūsoku-kyūsū	收敛级数 shoulian jishu, shoulian xilie
293	converted deviation	signal *m* de correction	umgeformte Regelabweichung	сигнал рассогласования, сигнал ошибки	deviazione convertita	señal *f* de error		转换偏差 zhuanhuan piancha
294	converted input signal	signal *m* d'entrée traduit	umgeformtes Eingangssignal	преобразованный входной сигнал	segnale d'ingresso convertito	señal *f* traducida de entrada	転換入力信号 tenkan-nyūryoku-shingo	转换输入信号 zhuanhuan shuru xinhao
295	converted output signal	signal *m* de sortie traduit	umgeformtes Ausgangssignal	преобразованный выходной сигнал	segnale di uscita convertito	señal *f* traducida de salida	転換出力信号 tenkan-shutsuryoku-shingo	转换输出信号 zhuanhuan shuchu xinhao
296	converted variable	grandeur *f* traduite	umgeformte Variable	преобразованная переменная	variabile convertita	variable *f* traducida	転換変量 tenkan-henryo	转换变量 zhuanhuan bianliang
297	converter	convertisseur *m*	Umformer	преобразователь	convertitore	conversor *m* convertidor *m*	変換器 コンバータ henkaki konbāta	变流器，转换器 bianliuqi, zhuanhuanqi
298	convolution integral	intégrale *f* de convolution	Faltungsintegral	интеграл свертки, интеграл Дюамеля	integrale di convoluzione	integral *f* de convolución	たたみこみ積分 tatamikomi-sekibun	卷积积分 juanji jifen
299	cooperative control	régulation *f* coopérative	gemeinsame Steuer- bzw. Regelung	согласованное управление	controllo cooperativo	control cooperativo	協調制御 kyōchō-seigyo	协作控制，合作控制 xiezuo kongzhi, hezuo kongzhi

	English	French	German	Russian	Italian	Spanish	Japanese	Chinese
300	coordinator	coordonnateur *m*	Koordinator	координатор, согласователь	coordinatore	coordinador	調整者 chōseisha	协调器 xietiaoqi
301	corner frequency	fréquence *f* de variation	Eckfrequenz	угловая частота	frequenza (o pulsazione) d'angolo	frecuencia *f* de inflexión	折点周波数 setten-shūhasū	角频率 jia pinlü
302	correcting range	étendue *f* de correction	Stellbereich, Korrekturbereich *	диапазон коррекции	campo di correzione	margen *m* de corrección	訂正範囲	校正范围 jiaozheng fanwei
303	correcting condition	condition *f* de correction	Korrekturbedingung, Stellgröße	условие коррекции	fase di correzione	condición *f* de corrección	訂正条件 teisei-jyōken	校正条件 jiaozheng tiaojian
304	correcting element	organe *m* correcteur	Korrekturglied, Stellglied	корректирующий элемент	elemento correttivo	elemento *m* corrector	訂正要素 teisei-yōso	校正环节 jiaozheng huanjie
305	correcting feedback	réaction *f* correctrice	korrigierende Rückführung	корректирующая обратная связь	retroazione correttiva	realimentación *f* correctora	訂正フィードバック teisei-fidobakku	校正反馈 jiaozheng fankui
306	correcting feedforward	action *f* directe correctrice	korrigierender Vorwärtszweig	корректирующая прямая связь	azione correttiva diretta	anticipación *f* correctora	訂正前饋 teisei-fidofowādo	校正前馈 jiaozheng qiankui
307	correcting range	étendue *f* de correction	Stellbereich, Korrekturbereich *	диапазон коррекции	campo di correzione	margen *m* de corrección	訂正動作範囲 teiseidōsa-han-i	校正范围 jiaozheng fanwei
308	correcting variable	grandeur *f* réglante	Korrekturgröße, Stellgröße	корректирующая величина	variabile correttiva	variable *f* correctora	操作量 sōsa-ryō	校正变量 jiaozheng bianliang
309	correction	correction *f*	Korrektur	коррекция	correzione	corrección *f*	訂正 teisei	校正 jiaozheng
310	correction time	durée *f* de réglage	Ausregelzeit	время восстановления	tempo di correzione	tiempo *m* de corrección	訂正時間 teisei-jikan	校正时间 jiaozheng shijian
311	corrective action	action *f* correctrice	Korrektureingriff, Korrekturvorgang	корректирующее воздействие	azione correttiva	acción *f* correctora	訂正動作 teisei-dōsa	校正作用 jianozheng zuoyong
312	correlation coefficient	coefficient *m* de corrélation	Korrelationskoeffizient	коэффициент корреляции	coefficiente di correlazione	coeficiente * de correlacion	相関係数 sōkan-keisū	相关系数 xiangguan xishu

24

No.	English	French	German	Russian	Italian	Spanish	Japanese	Chinese
313	Coulomb damping	amortissement _m_ de Coulomb	Coulombsche Dämpfung	демпфирование с сухим трением	smorzamento di Coulomb	amortiguamiento _m_ de coulomb	クーロン減衰 kūron-gensui	库仑阻尼 kulun zuni
314	Coulomb friction	frottement _m_ solide	Coulombsche Reibung	сухое трение	attrito di Coulomb	rozamiento _m_ de coulomb rozamiento _m_ seco	固体摩擦 kotai-masatsu	库仑摩擦, 干摩擦 kulun moca, gan moca
315	coupling	accouplement _m_, couplage _m_	Kopplung	соединение, связь	accoppiamento	acoplamiento _m_	干渉 kanshō	耦合 ouhe
316	coupling function	fonction _f_ de couplage	Kopplungsfunktion	функция связи	funzione di accoppiamento	función _f_ de acoplamiento	干渉関数 kanshō-kansū	耦合函数, 耦合功能 ouhe hanshu, ouhe gongneng
317	covariance matrix	matrice _f_ de covariance	Kovarianzmatrix	ковариационная матрица	matrice di covarianza	matriz _f_ covariante matriz _f_ de covarianza	共分散行列 kyōbunsan-gyōretsu	协方差矩阵 xiefangcha juzhen
318	criterion	critère _m_	Kriterium	критерий	criterio	criterio _m_	評価基準 基準 hyōka-kijun kijun	准则 zhunze
319	criterion function	fonction _f_ caractéristique, de critère	Gütefunktion	критерий управления	cifra di merito	funcón * criterio	評価関数 hyōka-kansū	准则函数, 判别函数 zhunze hanshu, panbie hanshu
320	critical damping	amortissement _m_ critique	kritische Dämpfung	критическое затухание	smorzamento critico	amortiguamiento _m_ crítico	臨界減衰 rinkai-gensui	临界阻尼 linjie zuni
321	critical path analysis	analyse _f_ des chemins critiques	Analyse des kritschen Pfades (Methode der Netzplantechnik)	анализ критического пути	analisi del cammino critico	análisis _m_ del camino crítico	クリティカルパス解析法 kuritikarupasu-kaisekihō	关键路径分析 guanjian lujing fenxi
322	critical point	point _m_ critique	kritischer Punkt	критическая точка	punto critico	punto _m_ critico	臨界点 rinkai-ten	临界点 linjiedian
323	cross compiler	compilateur _m_ de transfert	Kreuzkompilierer	кросскомпилятор	compilatore trasversale	compilador _m_ cruzado	クロスコンパイラ kurosu-konpaira	交叉编译程序 jiaocha bianyi chengxu
324	crosscorrelation	corrélation _f_ mutuelle	Kreuzkorrelation	взаимная корреляция	correlazione mutua	correlación _f_ cruzada	相互相関 sōgo-sōkan	互相关 huxiangguan
325	crosscorrelation function	function fonction _f_ de corrélation mutuelle	Kreuzkorrelations-funktion	взаимная корреляционная функция	funzione di correlazione mutua	función _f_ de correlación cruzada	相互相関関数 sōgo-sōkan-kansū	互相关函数 huxiangguan hanshu

	English	French	German	Russian	Italian	Spanish	Japanese	Chinese
326	crossover frequency	fréquence f de coupure	Übergangsfrequenz, Durchtrittsfrequenz	частота разделения, частота перехода, переходная частота	frequenza (o pulsazione) di attraversamento	frecuencia f de cruce	交点周波数 kōten-shūhasū	穿越频率 chuanyue pinlü
327	current amplifier	amplificateur m de courant	Stromverstärker	усилитель тока	amplificatore di corrente	amplificador m de corriente	電流増幅器 denryū-zōfukuki	电流放大器 dianliu fangdaqi
328	cursor	curseur m	Zeiger, Schreibzeiger, Positionsanzeiger	курсор	cursore	cursor m	カーソル kāsoru	光标、游标 guangbiao, youbiao
329	cushion	amortisseur m à fluide	dämpfen, abfedern, polstern	гидравлический амортизатор	cuscino, rivestimento elastico	amortiguador m hidráulico	緩衝器 kan-shō-ki	缓冲器 huanchongqi
330	cut-off frequency	fréquence f de coupure	Abschaltfrequenz	предельная частота, критическая частота, частота среза	frequenza (o pulsazione) d'interdizione	frecuencia f de corte	カットオフ周波数 しゃ断周波数 katto-ofu-syūhasū shadan-shūhasū	截止频率 jiezhi pinlü
331	cybernetics	cybernétique f	Kybernetik	кибернетика	cibernetica	cibernética f	サイバネティクス saibanetikasu	控制论 kongzhilun
332	cylinder	cylindre m	Zylinder	цилиндр	cilindro	cilindro	シリンダ shirinda	圆柱体 yuanzhuti
333	damped frequency	fréquence f naturelle amortie	Eigenfrequenz des gedämpften Systemes	собственная частота (затухающих колебаний)	frequenza (o pulsazione) smorzata	frecuencia f natural amortiguada pseudopulsación f	減衰周波数 gensui-shūhasū	阻尼频率 zuni pinlü
334	damped natural frequency	fréquence f naturelle amortie	gedämpfte Eigenfrequenz	собственная частота (затухающих колебаний)	frequenza (o pulsazione) naturale smorzata	frecuencia f natural amortiguada pseudopulsación f	減衰固有周波数 gensui-koyū-shūhasū	阻尼自然频率 zuni ziran pinlü
335	damped oscillation	oscillation f amortie	gedämpfte Schwingung	затухающее колебание	oscillazione smorzata	oscilación f amortiguada	減衰振動 gensui-shindō	阻尼振荡 zuni zhendang
336	damper	amortisseur m	Dämpfer	демпфер	smorzatore	amortiguador m	ダンパ 減衰器 danpa gensui-ki	阻尼器 zuniqi
337	damping	amortissement m	Dämpfung, Bedämpfung	демпфирование, затухание	smorzamento	amortiguamiento m	減衰 gensui	阻尼 zuni
338	damping coefficient	facteur m, coefficient m d'amortissement	Dämpfungskoeffizient	коэффициент демпфирования	coefficiente di smorzamento	coeficiente de amortiguamiento	減衰係数 gensui-keisū	阻尼系数 zuni xishu

No.	English	French	German	Russian	Italian	Spanish	Japanese	Chinese
339	damping constant	constante f d'amortissement	Dämpfungskonstante	коэффициент демпфирования	costante di smorzamento	constante f de amortiguamiento	減衰定数 gensui-jōsū	阻尼常数 zuni changshu
340	damping factor	facteur m, rapport m d'amortissement	Dämpfungsfaktor	коэффициент затухания	fattore di smorzamento	coeficiente m de amortiguamiento amortiguamiento m relativo	減衰係数 gensui-keisū	阻尼因子 zuni yinzi
341	damping ratio	taux m d'amortissement	Dämpfungsverhältnis	степень затухания	rapporto di smorzamento	razón f de amortiguamiento	減衰比 gensui-hi	阻尼比 zunibi
342	dashpot	amortisseur m à friction	Stoßdämpfer, Bremszylinder	амортизатор	smorzatore (ammortizzatore) a olio	amortiguador m	ダッシュポット dasshupotto	阻尼延迟器，缓冲器 zuni yanchiqi, huanchongqi
343	data acquisition	saisie f des données	Datenerfassung	сбор данных	acquisizione dei dati	adquisición f de datos	データ収集 dēta-shūshū	数据采集 shuju caiji
344	data handling (= processing)	traitement m de(s) données	Datenverarbeitung	обработка данных	elaborazione (di) dati	proceso de datos	データ処理 dēta-shori	数据处理 shuju chuli
345	data hold	bloqueur m, extrapolateur m	Halteglied, Datenspeicher	хранение данных	tenuta di un dato	retenedor	データ保持 dēta-hoji	数据保持 shuju baochi
346	data logger	collecteur m, enregistreur m d'informations	Datenerfassungseinrichtung	устройство записи данных	registratore di dati	registrador m (automático) de datos	データロガー dēta-rogā	数据记录装置 shuju jilu zhuangzhi
347	data logging	consignation f des informations	Daten erfassen (od. sammeln)	запись данных	registrazione di dati	registro m (automático) de datos	データロギング dēta-rogingu	数据记录 shuju jilu
348	data privacy	secret m des données	Datenschutz	приватность данных	riservatezza dei dati	privacidad * de datos		数据保密 shuju baomi
349	data processing	traitement m de l'information	Datenverarbeitung	обработка данных	elaborazione (di) dati	proceso m de datos	データ処理 dēta-shori	数据处理 shuju chuli
350	data processor	calculateur m	Datenverarbeitungsgerät	процессор обработки данных	elaboratore di dati	procesador m de datos	データプロセッサ dēta-purosessa	数据处理器 shuju chuliqi
351	data recorder	enregistreur m de données	Datenregistriergerät	устройство записи данных	registratore di dati	registrador * de datos	データ記録計 データ処理装置 dēta-kirokukei dēta-shori-sōchi	数据记录器 shuju jiluqi

#	English	French	German	Russian	Italian	Spanish	Japanese	Chinese
352	data reduction	compression f, réduction f des données	Datenreduktion	преобразование данных	riduzione dei dati	reducción * de datos	データ処理 dēta-shori	数据简化 shuju jianhua
353	data transmission	transmission f des données	Datenübertragung	передача данных	trasmissione di dati	transmisión * de datos	データ伝送 dēta-densō	数据传输 shuju chuanshu
354	database management	gestion f d'une banque de données	Datenbankverwaltung	управление базами данных	gestione di una base di dati	gestion * de bases de datos	データベース管理 dēta-bēsu-kanri	数据库管理 shujuku guanli
355	database structure	format m, structure f d'une banque de données	Datenbankstruktur	структура базы данных	struttura di una base di dati	estructura * de bases de datos	データベース構造 dēta-bēsu-kōzō	数据库结构 shujuku jiegou
356	dead band	seuil m de mobilité	tote Zone, Unempfindlichkeitsbereich	зона нечувствительности, "мёртвая" зона	banda soppressa	zona f muerta	不感帯 chūritsutai fudōtai	死区 siqu
357	dead band error	erreur f de mobilité	Unempfindlichkeitsfehler	ошибка зоны нечувствительности	errore di banda soppressa	error m de zona muerta	不感帯誤差 fukantai gosa	死区误差 siqu wucha
358	dead-beat response	réponse f pile	endliche Ausregelzeit	апериодическая реакция	risposta piatta	respuesta f amortiguada respuesta f aperiódica	有限時間整定応答 (デッドビート応答) yūgenjikan-seitei-ōtō (deddo-bīto-ōtō)	非周期响应 feizhouqi xiangying
359	dead time	temps m mort	Totzeit	запаздывание	tempo morto	tiempo m muerto	むだ時間 mudajikan	死时，停滞时间 sishi, tingzhi shijian
360	dead zone	zone f d'insensibilité	tote Zone	зона нечувствительности, "мёртвая" зона	zona morta	zona f muerta	中立帯 不動帯 chūritsutai fudōtai	死区 siqu
361	decentralized control	régulation f décentralisée	dezentrale Regelung od. Steuerung	децентрализованное управление	controllo decentralizzato	control m descentralizado	分散制御 bunsan-seigyo	分散控制 fensan kongzhi
362	decentralized controller	régulateur m décentralisé	dezentraler Regler	децентрализованный регулятор	controllore decentralizzato	controlador m descentralizado	分散型制御装置 bunsangata-seigyo-sōchi	分散控制器 fensan kongzhiqi
363	decentralized system	système m décentralisé	dezentralisiertes System	децентрализованная система	sistema decentralizzato	sistema m descentralizado	分散系 bunsan-kei	分散系统 fensan xitong
364	decibel	décibel m	Dezibel (dB)	децибел	decibel	decibelio m	デシベル deshiberu	分贝 fenbei

No.	English	French	German	Russian	Italian	Spanish	Japanese	Chinese
365	decision	décision m	Entscheidung	решение	decisione	decisión f	決定 kettei	决策 juece
366	decision table	table f de décision	Entscheidungstabelle	таблица решений	tabella di decisione	tabla f de decisión	決定表 shinnichi-hyō	决策表 juecebiao
367	decision theory	théorie f de la décision	Entscheidungstheorie	теория решений	teoria delle decisioni	teoría de la decisión	意志決定理論 ishi-kettei-riron	决策理论 juece lilun
368	decoder	décodeur m	Decodierer, Entschlüssler	декодер, декодирующее устройство	decodificatore	decodificador m	デコーダ 解読器 dekōda kaidokuki	译码器, 译码员 yimaqi, yimayuan
369	decoding	décodage m	dekodieren, entschlüsseln	декодирование	decodifica	decodificación f	復号 fukugō	译码 yima
370	decomposition	décomposition f	Dekomposition	декомпозиция	scomposizione	descomposición f	分割 bunkatsu	分解 fenjie
371	decoupled (or free) subsystems	sous-systèmes m.pl découplés	entkoppeltes (od. freies) Teilsystem	рассвязанные (=автономные) подсистемы	sottosistemi disaccoppiati (o liberi)	subsistemas m desacoplados m	非干渉部分システム hikanshō-buban-shisutema	解耦子系统, 去耦子系统 jie'ou zixitong, quou zixitong
372	decoupling	découplage m	Entkopplung	расцепление, рассвязка	disaccoppiamento	desacoplamiento m desacoplo m	非干渉化 hikanshōka	解耦 jie'ou
373	decoupling method	méthode f de découplage	Entkopplungsverfahren	метод рассвязки	metodo di disaccoppiamento	método m de desacoplo	非干渉法 hikanshō-hō	解耦方法 jie'ou fangfa
374	decrement	décrément m	Dekrement, Abnahme	декремент	decremento	decremento m	減衰率 gensui-ritsu	衰减, 衰减率 shuaijian, shuaijianlü
375	de-energize (v)	désexciter	entregen, abschalten, stromlos machen	снимать возбуждение, обесточивать	spegnere	desexcitar	励起を切る reiki-o-kiru	切断电源, 断电 qieduan dianyuan, duandian
376	definite-correction action	action f de correction déterminée	(definitive) Korrekturmaßnahme	определённое корректирующее воздействие	azione di correzione (a base) definita	acción correctora		限量校正作用 xianliang jiaozheng zuoyong
377	degenerative feedback	réaction f négative	Gegenkopplung (negative Rückkopplung)	отрицательная обратная связь	retroazione degenerativa	realimentación f degenerativa	負のフィードバック fu no fīdobakku	负反馈 fufankui

No.	English	French	German	Russian	Italian	Spanish	Japanese	Chinese
378	delay	retard *m*	Verzögerung	задержка, запаздывание	ritardo	retardo *m* retraso *m*	おくれ okure	延迟, 时延 yanchi, shiyan
379	delay element	organe *m* temporisateur	Verzögerungselement	элемент задержки	elemento di ritardo	elemento *m* de retardo	遅延要素 chien-yōso	延迟环节, 延迟元件 yanchi huanjie, yanchi yuanjian
380	demodulator	démodulateur *m*	Demodulator	демодулятор	demodulatore	demodulador *m*	デモジュレータ 復調器 demojurēta fukuchōki	解调器 jietiaoqi
381	derivative action	action *f* par dérivation, action *f* D	D-Verhalten (differenzierender Einfluß), Vorhalt	воздействие по скорости, воздействие по производной	azione derivativa	acción *f* derivada acción *f* D	微分動作 bibundōsa	微分作用 weifen zuoyong
382	derivative action factor	coefficient *m* d'action par dérivation	Differenzierbeiwert	коэффициент воздействия по производной	fattore (o coefficiente) dell'azione derivativa	coeficiente *m* de acción derivada	微分動作係数 bibun-dōsakeisū	微分因子 weifen yinzi
383	derivative action time	temps *m* (de dosage) de dérivation, temps *m* d'activation	Vorhaltzeit	время предварения	tempo dell'azione derivativa	tiempo *m* de acción derivada tiempo *m* de anticipación	微分動作時間 bibun-dōsajikan	微分时间 weifen shijian
384	derivative element	organe *m* d'action par dérivation	Differenzierglied	дифференцирующий элемент, дифференциатор	elemento derivativo	elemento *m* derivador	微分要素 bibun-yōso	微分环节 weifen huanjie
385	describing function	transmittance *f* isochrone équivalente	Beschreibungsfunktion	описывающая функция	funzione descrittiva	función *f* descriptiva	記述関数 kijutsu-kansū	描述函数 miaoshu hanshu
386	detectability	décelabilité *f*	Erfaßbarkeit, Nachweisbarkeit, Erkennbarkeit	обнаруживаемость, детектируемость	rivelabilità	detectabilidad *f*	可検出性 ka-kenshutsusei	可检测性 kejiancexing
387	detecting element	capteur *m*	Erfassungsgerät, Meßfühler	детектор	elemento rivelatore	elemento *m* detector	検出要素 kenshutsu-yōso	检测环节, 检测元件 jiance huanjie, jiance yuanjian
388	detector	capteur *m*	Fühler, Meßwertgeber, Detektor	детектор	rivelatore	detector *m*	検出器 kenshutsu-ki	检出器 jianchuqi
389	deterministic	déterministe	deterministisch	детерминированный	deterministico	determinístico	確定論的 kakuteironteki	确定的 quedingde
390	deterministic system	système *m* déterministe	deterministisches System	детерминированная система	sistema deterministico	sistema *m* determinístico	確定論的システム kakuteironteki-shisutemu	确定性系统 quedingxing xitong

No.	English	French	German	Russian	Italian	Spanish	Japanese	Chinese
391	deviation	écart m	Abweichung, Sollwertabweichung	отклонение	deviazione	desviación f	偏差 hensa	偏差 piancha
392	diagnostic program	programme m de recherche d'erreurs de programmation	Diagnoseprogramm	диагностическая программа	programma diagnostico	programas * de diagnóstico	プログラム診断 purogurama-shindan	诊断程序 zhenduan chengxu
393	diagnostic test	test m diagnostique	Diagnoseuntersuchung	диагностический тест	prova orientata alla diagnosi	test m de diagnóstico	診断 shindan	诊断检测 zhenduan jiance
394	diagonal dominance	effet m diagonal dominant	Diagonaldominanz	диагональное доминирование	dominanza diagonale	predominio m diagonal	対角優勢 taikakuyūsei	对角主导 duijiao zhudao
395	diagonal matrix	diagonale f d'une matrice	Diagonalmatrix	диагональная матрица	matrice diagonale	matriz * diagonal	対角行列 taikaku-gyōretsu	对角矩阵 duijiao juzhen
396	diagram	schéma m, diagramme m	Diagramm, graphische Darstellung, Schaltplan	граф, схема, диаграмма	diagramma	diagrama m esquema m	線図 senzu	图，图表 tu, tubiao
397	diaphragm	membrane f	Membrane, Blende	диафрагма, мембрана	diaframma	membrana f	ダイアフラム daiafuramu	膜片，隔膜 mopian, gemo
398	diaphragm actuator	servomoteur m à membrane	Membranantrieb	мембранный привод	attuatore a diaframma	accionador m de mambrana	ダイアフラム アクチュエータ daiafuramu-akuchuēta	膜片执行机构 mopian zhixing jigou
399	diaphragm valve	vanne f à membrane	Membranventil	мембранный клапан	valvola a diaframma	válvula f de membrana	ダイアフラム弁 daiafuramu-ben	隔膜阀 gemofa
400	difference equation	équation f aux différences	Differenzengleichung	разностное уравнение	equazione alle differenze	eduación f en diferencias	差分方程式 sabun-hōteishiki	差分方程 chafen fangcheng
401	differential amplifier	amplificateur m différentiel	Differentialverstärker	дифференциальный усилитель	amplificatore differenziale	amplificador m diferencial	差動増幅器 sado-zōfukuki	差动放大器，微分放大器 chadong fangdaqi, weifen fangdaqi
402	differential analyser	analyseur m différentiel	Differentialanalysator	дифференциальный анализатор	analizzatore differenziale	analizador m diferencial	微分解析機 bibun-kaisekiki	差动分析器 chadong fenxiqi
403	differential equation	équation f différentielle	Differentialgleichung	дифференциальное уравнение	equazione differenziale	ecuación * diferencial	微分方程式 bibun-hōteishiki	微分方程 weifen fangcheng

No.	English	French	German	Russian	Italian	Spanish	Japanese	Chinese
404	differential field motor	moteur *m* à excitations inverses	Gegencompoundmotor	двигатель с расщепленными полюсами	motore a campo differenziale	motor *m* de campo partido		裂磁式电动机 liechangshi diandongji
405	differential gap	saut *m* différentiel	Schaltdifferenz (zwischen oberem und unterem Umschaltwert)	нерегулируемый интервал, нейтральная зона	intervallo differenziale	intervalo *m* diferencial	動作すきま dōsa-sukima	差动间隙 chadong jianxi
406	differential gear	différentiel *m*	Differentialgetriebe	дифференциальная передача	(ingranaggio) differenziale	marcha diferencial	差動歯車 sadō-haguruma	差动齿轮 chadong chilun
407	differential lever	levier *m* différentiel	Differentialhebel	маятник	leva differenziale	palanca diferencial	差動てこ sadō-teko	差动杆 chadonggan
408	differential transfer factor	factfacteur *m* de transfert différentiel	Differential-Übertragungsfaktor	дифференциальный коэффициент передачи	fattore di trasferimento differenziale	factor de transferencid diferencial		差动传递因子 chadong chuandi yinzi
409	differential transformer	transformateur-comparateur *m*	Differentialübertrager	линейный дифференциальный трансформатор	trasformatore differenziale	transformador *m* diferencial	差動変圧器 sadō-hen-atsuki	差接变压器 chajie bianyaqi
410	differentiating action	action *f* dérivatrice	differenzierender Einfluß	дифференцирующее воздействие	azione derivatrice	acción *f* diferenciadora	微分動作 bibun-dōsa	微分作用 weifen zuoyong
411	differentiating element	organe *m* différentiateur	differenzierendes Element	дифференциатор	elemento derivatore	elemento *m* diferenciador	微分要素 bibun-yōso	微分环节，微分元件 weifen huanjie, weifen yuanjian
412	differentiator	organe *m* différentiateur	Differenzierschaltung, Differenziergerät	дифференциатор	differenziatore	diferenciador *m*	微分器 bibun-ki	微分器 weifenqi
413	digital-analog converter	convertisseur *m* numérique-analogique	Digital-Analog-(DA)-Wandler, (od. Umsetzer)	цифро-аналоговый преобразователь (ЦАП)	convertitore digitale-analogico	conversor *m* digital-analógico, convertidor *m* digital-analógico	ディジタル-アナログ変換器 dijitaru-anarogu-henkanki	数模转换器 shumo zhuanhuanqi
414	digital circuit	circuit *m* numérique	Digitalschaltung	цифровая цепь	circuito digitale	circuito digital	ディジタル回路 dijitaru-kairo	数字线路，数字电路 shuzi xianlu, shuzi dianlu
415	digital computer	calculateur *m* numérique, calculatrice *f* numérique	Digitalrechner	ЭВМ, цифровая вычислительная машина	calcolatore digitale	calculador *m* digital, computador *m* digital	ディジタル計算機 dijitaru-keisanki	数字计算机 shuzi jisuanji
416	digital control	commande *f* numérique	Digitalregelung, digitale Steuerung	цифровое управление	controllo digitale	control *m* digital	ディジタル制御 dijitaru-seigyo	数字控制 shuzi kongzhi

No.	English	French	German	Russian	Italian	Spanish	Japanese	Chinese
417	digital differential analyser	analyseur m différentiel numérique	digitaler Differenzensummator (DDA)	цифровой дифференциальный анализатор	analizzatore differenziale digitale	analizador m diferencial digital	ディジタル微分解析機 dijitaru-bibun-kaisekiki	数字微分分析器 shuzi weifen fenxiqi
418	digital filter	filtre m numérique	Digitalfilter	цифровой фильтр	filtro digitale	filtro m digital	ディジタルフィルタ dijitaru-firuta	数字滤波器 shuzi lüboqi
419	digital image	image f numérique	digitalisiertes Bild	цифровое изображение	immagine digitale	imagen digital	ディジタル像 dijitaru-zō	数字图象 shuzi tuxiang
420	digital signal	signal m numérique	Digitalsignal	цифровой сигнал	segnale digitale	señal f digital	ディジタル信号 dijitaru-shingō	数字信号 shuzi xinhao
421	digital system	système m discret, numérique	Digitalsystem	цифровая система	sistema digitale	sistema * digital	ディジタルシステム dijitaru-shisutemu	数字系统 shuzi xitong
422	digitizer	codeur m numérique	Digitalisierer, Digitalwandler	квантующее устройство	digitizzatore	codificador m numérico digitizador m	ディジタイザ dijitaiza	数字化仪 shuzihua yi
423	dimension (of a vector space)	dimension f (d'un espace vectoriel)	Dimension (eines Vektorraumes)	размерность (векторного пространства)	dimensione (di uno spazio vettoriale)	dimensión f (de un espacio vectorial)	次元 jigen	(向量空间的)维数 (xiangliang kongjiande) weishu
424	diode function generator	générateur m de fonction à diodes	Diodenfunktionsgenerator	функциональный преобразователь на диодах	generatore di funzioni a diodi	generador m de función de diodos	ダイオード関数発生器 daiōdo-kansū-hasseiki	二极管函数发生器 erjiguan hanshu fashengqi
425	Dirac function	fonction f de Dirac	Dirac-Funktion, d-Funktion	единичный импульс, дельта импульс	funzione di Dirac	función f de Dirac	デルタ関数 deruta-kansū	狄拉克函数 dilake hanshu
426	direct digital control	commande f numérique directe	direkte digitale Regelung	прямое цифровое управление	controllo digitale diretto	control m digital directo	直接ディジタル制御 chokusetsu-dijitaru-seigyo	直接数字控制 zhijie shuzi kongzhi
427	directed graph	graphe m orienté	gerichteter Graph	ориентированный граф (=орграф)	grafo orientato	grafo m orientado	有向グラフ yūko-gurafu	有向图 youxiangtu
428	discontinuous action	action f intermittente, discontinue	unstetige Beeinflussung (Regler)	прерывистое	azione discontinua	acción f discontinua acción f intermitente	不連続動作 furenzoku-dōsa	不连续作用 bulianxu zuoyong
429	discontinuous-action controller	régulateur m à action discontinue	unstetiger Regler	Iвоздействие тульсного действия	controllore ad azione discontinua	controlador m de acción discontinua controlador m de acción intermitente	不連続作動形調節器 furenzokudōsa-chōsetsuki	不连续作用控制器 bulianxu zuoyong kongzhiqi

№	English	Français	Deutsch	Русский	Italiano	Español	日本語	中文
430	discontinuous control	commande f discontinue, régulation f discontinue	unstetige Regelung od. Steuerung	прерывистое управление (регулирование)	controllo discontinuo	control m discontinuo, control m intermitente	不連続制御 furenzoku-seigyo	不连续控制 bulianxu kongzhi
431	discontinuous variable	grandeur f discontinue	unstetige (nichtstetige) Variable	прерывистая переменная	variabile discontinua	variable f discontinua	不連続変数 furenzoku-hensū	不连续变量 bulianxu bianliang
432	discrete system	système m discret	diskretes System	дискретная система	sistema discreto	sistema m discreto	離散系 risan-kei	离散系统 lisan xitong
433	discrete valued parameter	paramètre m à valeurs discrètes	(zeit-)diskreter Parameterwert	дискретный параметр	parametro a valori discreti	parámetro m de valores discretos	離散値パラメータ risanchi-paraméta	离散值参数 lisanzhi canshu
434	discrimination	résolution f	Unterscheidung, Selektivität	разрешающая способность	discriminazione	discriminación f	弁別 benbetsu	判别 panbie
435	disjunction	disjonction f	Disjunktion, inklusives ODER	дизъюнкция, логическая сумма	disgiunzione	disyunción f	論理和 risetsu	折取，"或" xiqu, "huo"
436	disk memory	mémoire f disque	Plattenspeicher	диск, накопитель на магнитном диске (НМД)	memoria a dischi	memoria f de disco	ディスクメモリ disuku-memori	磁盘存储器 cipan cunchuqi
437	distance/velocity lag	retard m de parcours	Weg-/Geschwindigkeits-verzögerung Nacheilung	время выбега (по положению)	ritardo (espresso dal rapporto) distanza/velocità	retardo de recorrido/velocidad	むだ時間 mudajikan	距离／速度滞后 juli/sudu zhihou
438	distortion	distorsion f	Verzerrung	искажение	distorsione	distorsión f	歪 hizumi	畸变，失真 jibian, shizhen
439	distributed computer control system (DCCS)	système m de réglage à calculateurs répartis (S.R.C.R.)	verteilte (Prozeßrechner-systeme)	распределенная система компьютерного управления	sistema di controllo distribuito mediante calcolatore	sistema m distribuido de control por computador, sistema m repartido de control por computador	分散型計算機制御系 bunsangata-keisanki-seigyo-kei	分布式计算机控制系统 fenbushi jisuanji kongzhi xitong
440	distributed control	régulation f répartie	verteilte od. dezentrale Regelung (Steuerung)	распределенное управление	controllo distribuito	control m distribuido, control m repartido	分散制御 bunsan-seigyo	分布控制 fenbu kongzhi
441	distributed model	modèle m réparti	verteiltparametrisches od. dezentrales Modell	модель с распределенными параметрами	modello distribuito	modelo m distribuido, modelo m repartido	分布モデル bunpu-moderu	分布模型 fenbu moxing
442	distributed parameter	paramètre m réparti	verteilte Parameter	распределенный параметр	a parametri distribuiti	parámetros m distribuidos, parámetros m repartidos	分布パラメータ bunpu-paraméta	分布参数 fenbu canshu

No.	English	French	German	Russian	Italian	Spanish	Japanese	Chinese
443	distributed-parameter system	système *m* à paramètres répartis	verteiltparametrisches System	система с распределёнными параметрами	sistema a parametri distribuiti	sistema *m* de parámetros distribuidos, sistema *m* de parámetros repartidos	分布パラメータ系 bunpu-paramēta-kei	分布参数系统 fenbu canshu xitong
444	distributed system	système *m* réparti	verteiltes od. dezentralisiertes System	распределённая система	sistema distribuito	sistema *m* distribuido, sistema *m* repartido	分布系 bunpu-kei	分布系统 fenbu xitong
445	disturbance	perturbation *f*	Störung, Störgröße	возмущение	disturbo	perturbación *f*	外乱 gairan	扰动 raodong
446	disturbance parameter	paramètre *m* de perturbation	Störparameter	параметр возмущения	disturbo parametrico	parámetro *m* de perturbación	外乱パラメータ gairan-paramēta	扰动参数 raodong canshu
447	disturbance rejection	réjection *f* des perturbations	Störunterdrückung	подавление помехи, подавление возмущения	reiezione del disturbo	rechazo *m* de perturbación	外乱除去 gairan-jyokyo	扰动抑制 raodong yizhi
448	disturbance signal	signal *m* perturbateur	Störsignal	возмущающее воздействие	(segnale di) disturbo	señal *f* perturbadora	外乱信号 gairan-shingō	扰动信号 raodong xinhao
449	disturbance variable	grandeur *f* perturbatrice	Störgröße	переменная возмущения	(variabile di) disturbo	variable *f* perturbadora	外乱 gairan	扰动变量 raodong bianliang
450	disturbance-variable compensation	compensation *f* de perturbation	Störgrößenkompensation	компенсация возмущений	compensazione del disturbo	compensación *f* de la variable perturbadora	外乱補償 gairan-hoshō	扰动变量补偿 raodong bianliang buchang
451	dither	titillation *f*	Zitterbewegung, zittern	дрожание	sovreccitazione	temblor *m*	ディザ diza	高频颤动 gaopin chandong
452	divided monitoring feedback	réaction *f* principale partagée	getrennte Kontrollrückführung od. stabilisierende Rückführung	разделённая контролирующая обратная связь	supervisore in retroazione parcellizzato	realimentación *f* de monitorización dividida	分割監視フィードバック bunkatsu-kanshi-fidobakku	分割监控反馈 fenlu jiankong fankui
453	divider	diviseur *m*	Dividierer, Teiler, Spannungsteiler	делитель	divisore (partitore)	divisor *m*	除算器 jozankikō	除法器，分压器 chufaqi, fenyaqi
454	dog clutch	clabot *m*	Antriebsklaue, Antriebskupplung, Schaltklaue	кулачковая муфта	giunto (o innesto) a denti	embrague	ドグクラッチ dogu-kuratti	爪形离合器 zhaoxing liheqi
455	dominant roots	racines *f pl* dominantes	dominante Wurzeln	доминантные корни	radici dominanti	raices *f* dominantes	卓越根 takuetsu-kon	主根 zhugen

No.	English	French	German	Russian	Italian	Spanish	Japanese	Chinese
456	doublet impulse	doublet m impulsionnel	Doppelimpuls	сдвоенный импульс	doppio impulso	doblete m	2重インパルス ni-jū-inparusu	双重线脉冲 shuangchongxian maichong
457	downtime	durée f d'indisponibilité, d'immobilisation, de défaillance	Ausfallzeit	время остановки, время нераработоспособности	tempo passivo	tiempo de parada	動作不能時間 dōsa-funō-jikan	停机时间，故障时间 tingji shijian, guzhang shijian
458	downward compatibility	compatibilité f avec l'aval	(Ausfall-) Kompatibilität	совместимость (сверху) вниз	compatibilità verso il basso	compatibilidad f descendente	下位両立性 下位適合性 kai-ryōritsusei kai-tekigōsei	向下兼容性 xiangxia jianrongxing
459	drag-cup motor	moteur m à cloche	Anwurfmotor	двигатель с полым ротором	motore a tazza	motor m de campana	ドラグカップ形電動機 doragukappugata-dendōki	(拖)杯式电动机 (tuo) beishi diandongji
460	drift	dérive f	Drift, Abwanderung, Auswanderung	дрейф	deriva	deriva f	ドリフト dorifuto	漂移 piaoyi
461	drift rate	taux m de dérive	Driftrate oder Auswanderung, Auswanderungsgeschwindigkeit (Kreisel)	скорость дрейфа	velocità di deriva	velocidad f de deriva	ドリフト速度 dorifuto-sokudo	漂移率 piaoyilü
462	droop	statisme m	Absenkung, Wölbung der Vorderkante	статизм	incurvamento	regulación f de carga, estatismo m		下垂度，倾斜 xiachui du, qingxie
463	drum memory	tambour m	Trommelspeicher	барабан, накопитель на магнитном барабане (НМБ)	memoria a cilindro	memoria f de tambor	ドラムメモリ doramu-memori	磁鼓存储器 cigushi cynchuqi
464	dry friction	frottement m solide	trockene Reibung	сухое трение	attrito secco	rozamiento m seco	固体摩擦 kansei-masatsu	干摩擦 ganmoca
465	dual-composition control	régulation f à compositions duelles	zweifach wirkende Regelung (Steuerung)	дуальное композиционное управление	controllo a composizione duale	control de composición dual	二成分制御 ni-seben-seigyo	双组合控制 shuangzuhe kongzhi
466	dual computer system	système m de calcul duel	Doppelrechnersystem	сдвоенная вычислительная система	sistema a calcolatore duale	sistema m de doble computador	2重計算系システム ni-jūjisuanji kongzhi	双计算机控制 shuangjisuanji kongzhi
467	dual-mode control	commande f à deux modes de fonctionnement	Zweipunktregelung	двухрежимное управление	controllo duale	control m biomodal	2重モード制御 nijyūmōdo-seigyo	双模式控制，双模态控制 shuangmoshi kongzhi, shuangmotai kongzhi
468	dual mode system	système m à effet duel	zweistufiges System, Zweipunktsystem	двухрежимная система	sistema a funzionamento duale	sistema m bimodal	2モード系 ni-mōdo-kei	双模式系统，双模态系统 shuangmoshi xitong, shuangmotai xitong

	English	French	German	Russian	Italian	Spanish	Japanese	Chinese
469	duality	dualité f	Dualismus, Dualität	двойственность	dualità	dualidad f	双対性 sōtsuisei	对偶性，二重性 duì'ouxing, erchongxing
470	Duhamel integral	intégrale f de Duhamel	Duhamelsches Integral	интеграл свертки, интеграл Дюамеля	integrale di Duhamel	integral f de Duhamel	デュハメルの積分 Dyuhameru-no-sekibun	卷积积分 juanji jifen
471	duplex control	commandes f pl jumelées	Duplexsteuerung	дуплексное управление	controllo duplex	control m de doble efecto, control m dúplex	複合制御 fukugō-seigyo	双工控制 shuanggong kongzhi
472	duty	utilisation f	Betriebsart, Betrieb, Auslastung	рабочий режим	compito	servicio m, trabajo m	デューティ dyūti	负荷状态，工作状态 fuhe zhuangtai, gongzuo zhuangtai
473	duty cycle	cycle m d'utilisation	Betriebsspeicher, Arbeitszyklus, Testverhältnis	рабочий цикл	ciclo di funzionamento, fattore di utilizzazione	ciclo m de trabajo	デューティサイクル dyūti-saikuru	工作循环 gongzuo xunhuan
474	duty factor	facteur m d'utilisation	Arbeitphase, Tastverhältnis (Impulstechnik)	коэффициент использования	fattore di utilizzazione	coeficiente m de trabajo, factor m de trabajo	デューティ因子 dyūti-inshi	占空因数，占空比 zhankong yinshu, zhankongbi
475	duty factor action	action f par le facteur d'utilisation	Tastverhältnisauswirkung	воздействие по коэффициенту использования	azione intermittente	acción de coeficiente de trabajo		占空因数作用 zhankong yinshu zuoyong
476	dynamic	dynamique, transitoire	dynamisch	динамический	dinamico	dinámico	動的な dōteki-na	动态的，动力学的 dongtaide, donglixuede
477	dynamic decoupling	découplage m transitoire	dynamische Entkopplung	динамическая развязка	disaccoppiamento dinamico	desacoplo m dinámico	動的非干渉化 dōteki-hikanshōka	动态解耦 dongtai jie'ou
478	dynamic programing	programmation f dynamique	dynamische Programmierung	динамическое программирование	programmazione dinamica	programación f dinámica	動的計画法 dōteki-keikaku-hō	动态规划 dongtai guihua
479	dynamic test	test m transitoire	dynamischer Test	динамический тест	prova dinamica	test m dinámico (verificación dinámica)	動的試験 dōteki-shiken	动态测试 dongtai ceshi
480	eddy-current brake	frein m à courant de Foucault	Wirbelstrombremse	тормоз на принципе вихревых токов	freno a corrente parassita	freno de corriente eddy	うず電流ブレーキ uzudennyū-burēki	涡流制动 woliu zhidong
481	effective dead time	retard m à la montée	effektive Totzeit	время задержки, время запаздывания	tempo morto effettivo	retardo m inicial, tiempo m muerto	有効むだ時間 yūkō-mudajikan	有效空载时间，有效死时 youxiao kongzai shijian, youxiao sishi

	English	French	German	Russian	Italian	Spanish	Japanese	Chinese
482	effective range	étendue f de mesure	effektiver Bereich	эффективный диапазон	gamma effettiva	campo m de medida	有効範囲 yūkō-han-i	有效范围 youxiao fanwei
483	eigenfunction	fonction f propre	Eigenfunktion	собственная функция	autofunzione	función f propia	固有関数 koyū-kansū	本征函数 benzheng hanshu
484	eigenvalue	valeur f propre	Eigenwert	собственное значение	autovalore	autovalor, valor m característico, valor m propio	固有値 koyū-chi	本征值，特征值 benzheng zhi, tezheng zhi
485	eigenvector	vecteur m propre	Eigenvektor	собственный вектор	autovettore	vector m propio	固有ベクトル koyū-bekutoru	本征向量 benzheng xiangliang
486	electrical angle	angle m électrique	elektrischer Winkel, Phasenwinkel	электрический угол	angolo elettrico	ángulo m eléctrico	電気角 denki-kaku	电角度 dian jiaodu
487	element	organe m, élément m	Element, Bauteil, Glied	элемент	elemento	elemento m	要素 yōso	环节，元件 huanjie, yuanjian
488	embedded system	système m caché	eingebettetes (auch math.) od. eingebautes System	вложенная (компьютерная) система	sistema incorporato	sistema empotrado		嵌入系统 qianru xitong
489	emitter follower	émettodyne f	Emitterfolger, Impedanzwandler	эмиттерный повторитель	inseguitore a emettitore comune	seguidor m de emisor	エミッタホロア emitta-horoa	射极跟随器 sheji gensuiqi
490	emulator	émulateur m	Emulator, Kompatibilitätsein-richtung	эмулятор	emulatore	emulador m	エミュレータ emyureita	仿真器，仿真程序 fangzhenqi, fangzhen chengxu
491	encoder	codeur m	Codierer, Verschlüßler, Drehgeber	кодирующее устройство	codificatore, trasduttore digitale di posizione (angolare)	codificador m	符号器 fugōki	编码器 bianmaqi
492	end-point control	régulation f asymptotique	Endwertregelung	терминальное управление	controllo terminale	control m de valor final	終点制御 shūten-seigyo	终端控制 zhongduan kongzhi
493	energize (v)	dynamiser (vt), alimenter (v)	erregen, einschalten	возбуждать	accendere	excitar	励起する reiki-suru	激励，通电 (动词) jili, tongdian (dongci)
494	envelope	enveloppe f	Einhüllende, Hüllkurve	огибающая	busta, inviluppo	envolvente f	包絡線 hōrakusen	包络 baoluo

#	English	French	German	Russian	Italian	Spanish	Japanese	Chinese
495	environmental coefficient	coefficient *m* de stabilité	Umweltkoeffizient	коэффициент стабильности	coefficiente ambientale	coeficiente *m* ambiental	等化 tōka	环境系数 huanjing xishu
496	environmental stability	stabilité *f* d'ambiance	Umgebungsstabilität	устойчивость к внешним воздействиям	stabilità ambientale	estabilidad en el entorno	対環境安定性 tai-kankyō-anteisei	环境稳定性 huanjing wendingxing
497	equalization	correction *f*, compensation *f*	Ausgleich bewirken	выравнивание	equalizzazione	ecualización *f*, compensación *f*	等化 tōka	补偿 buchang
498	equalizer	correcteur *m*, compensateur *m*	Kompensationsglied, Ausgleichseinrichtung	выравниватель	equalizzatore	ecualizador *m*, compensador *m*	等化器 tōkaki	补偿器, 补偿电路 buchangqi, buchangdianlu
499	equilibrium	équilibre *m*	Gleichgewichtslage, Gleichgewicht	равновесие	equilibrio	equilibrio *m*	平衡 heikō	平衡 pingheng
500	erect (v)	monter (v)	einrichten, aufstellen, montieren	монтировать, собирать	erigere	lanzar	起立する kiritsu-suru	安装、竖立（动词） anzhuang, shuli (dongci)
501	erection torque motor	moteur *m*	Aufrichtmotor	моментный мотор	motore coppia d'erezione	motor *m* de par de lanzamiento		竖起力矩马达 shuqi liju mada
502	ergodic theorem	théorème *m* d'ergodicité	Egodentheorem	эргодическая теорема	teorema ergodico	teorema *m* ergódico	エルゴードの定理 erugōdo-no-teiri	遍历定理 bianlixing dingli
503	error	erreur *f*	Fehler, Abweichung, Ungenauigkeit	погрешность, ошибка, сбой	errore	error *m*	誤差 偏差 gosa hensa	误差 wucha
504	error compensation	compensation *f* d'erreur	Fehlerkompensation	компенсация ошибки	compensazione dell'errore	compensación *f* de error	誤差補償 gosa-hoshō	误差补偿 wucha buchang
505	error signal	signal *m* de correction	Fehlersignal	сигнал рассогласования, сигнал ошибки	segnale errore	señal *f* de error	偏差信号 hensa-shingō	误差信号 wucha xinhao
506	error transfer function	transmittance *f* de la chaîne d'action	Fehlerübertragungs-funktion	передаточная функция сигнала ошибки	funzione di trasferimento all'errore	función *f* de transferencia de error	誤差伝達関数 gosa-dentatsu-kansū	误差传递函数 wucha chuandi hanshu
507	estimation	estimation *f*	Schätzung, Zustandsschätzung	оценивание	stima	estimación *f*	推定 suitei	估计 guji

No.	English	French	German	Russian	Italian	Spanish	Japanese	Chinese
508	estimation of parameters	estimation f paramétrique	Parameterschätzung	оценивание параметров	stima di parametri	estimación f de parámetros	パラメータ推定 parameta-suitei	参数估计 canshu guji
509	estimation of state	estimation f d'état	Zustandsschätzung	оценивание состояния	stima dello stato	estimación f del estado	状態推定 jōtai-suitei	状态估计 zhuangtai guji
510	excitation winding	enroulement m d'excitation	Erregerwicklung, Feldwicklung	обмотка возбуждения	avvolgimento di eccitazione	arrollamiento m de excitación	励磁巻線 reiji-makisen	励磁绕组 jici raozu
511	exciter	excitatrice f	Erreger, Erregermaschine	возбудитель	eccitatore	excitatriz f	励磁機 reijiki	励磁机, 激励器 liciji, liciqi
512	exclusion	exclusion f	Ausschließung, Inhibition	исключение	esclusione	exclusión f	排反 haihan	排斥, 不相容 paichi, buxiangrong
513	expectation	espérance f, attente f	Erwartung (statistisch)	ожидание	attesa	esperanza	期待値 kitai-chi	期望 qiwang
514	expected value	valeur f attendue	Erwartungswert	ожидаемое значение	valore atteso	valor m esperado	期待値 kitaichi	期望值 qiwangzhi
515	expert system	système m expert	Expertensystem	экспертная система	sistema esperto	sistemas * experto	エキスパートシステム ekisupāto-shisutemu	专家系统 zhuanjia xitong
516	exponential lag	retard m exponentiel	exponentielle Verzögerung	замедление, инертное замедление	ritardo esponenziale	retardo m exponencial	指数関数おくれ shisūkansū-okure	指数滞后 zhishu zhihou
517	exponentially stable	exponentiellement stable	exponentielle Stabilität	экспоненциально устойчивый	esponenzialmente stabile	exponencialmente estable	指数関数的安定 shisūkansūteki-antei	指数稳定 zhishuxing wending
518	external memory	mémoire f externe	externer Speicher	внешняя память	memoria esterna	memoria f externa	外部記憶 gaibu-kioku	外存储器 waicunchuqi
519	fail-soft	à dégradation progressive	Notfahrprogramm	мягкая деградация	a prova di malfunzionamento	resistente a fallos	フェイルソフト feirusofuto	故障弱化 guzhang ruohua
520	failure	défaillance f, panne f	Ausfall	повреждение, неисправность, отказ	fallimento, guasto	fallo m	故障 koshō	故障 guzhang

No.	English	French	German	Russian	Italian	Spanish	Japanese	Chinese
521	fast Fourier transform	transformée f de Fourier rapide	schnelle Fouriertransformation	быстрое преобразование Фурье	trasformata rapida di Fourier	transformada f rápida de Fourier	高速フーリェ変換 kōsoku-fūrie-henkan	快速傅立叶变换 kuaisu fuliye bianhuan
522	fault detection	détection f de défauts	Fehlererkennung	обнаружение неисправности	rivelazione di guasto	detección f de fallos	故障診断 koshō-shindan	故障探测 guzhang tance
523	fault location	localisation f de défauts	Fehlerort	локализация неисправности	individuazione di un guasto	localización f de fallos	故障位置 koshō-ichi	故障定位 guzhang dingwei
524	feedback	réaction f, rétroaction f	Rückführung	обратная связь	retroazione	realimentacion, retroalimentación f, retroacción f	帰還 fido-bakku	反馈 fankui
525	feedback amplifier	amplificateur m à réaction	rückgekoppelter Verstärker	усилитель обратной связи	amplificatore reazionato	amplificador m realimentado	フィードバック増幅器 fidobakku-zōfukuki	反馈放大器 fankui fangdaqi
526	feedback channel	chaîne f de réaction	Rückführzweig	канал обратной связи	canale di retroazione	cadena f de realimentación, canal m de realimentación	フィードバックチャネル fidobakku-chaneru	反馈信道, 反馈通道 fankui xindao, fankui tongdao
527	feedback control	régulation f par rétroaction	Regelung	управление с обратной связью	controllo a retroazione	control m con realimentación, control m realimentación	フィードバック制御 fidobakku-seigyo	反馈控制 fankui kongzhi
528	feedback control system	système m de régulation par rétroaction	Regelungssystem	система управления (или регулирования) с обратной связью	sistema di controllo a retroazione	sistema m de control con realimentación, sistema m realimentado de control	フィードバック制御系 fidobakku-seigyo-kei	反馈控制系统 fankui kongzhi xitong
529	feedback element	organe m de réaction	Glieder im Rückführsystem, Rückführelement	элемент обратной связи	elemento in retroazione	elemento m de realimentación	フィードバック要素 fidobakku-yōso	反馈环节 fankui huanjie
530	feedback loop	boucle f de réaction	Regelkreis, Rückkopplungsschleife	контур обратной связи	anello di retroazione	anillo m de realimentación, bucle m de realimentación, lazo m de realimentación	フィードバックループ fidobakku-rūpu	反馈回路 fankui huilu
531	feedback path	chaîne f de réaction	Rückführzweig, Rückführpfad	канал обратной связи	percorso di retroazione	camino m de realimentación	フィードバック径路 fidobakku-keiro	反馈通道 fankui tongdao
532	feedback signal	signal m de réaction	Rückführsignal	сигнал обратной связи	segnale di retroazione	señal f de realimentación	フィードバック信号 fidobakku-shingō	反馈信号 fankui xinhao
533	feedback variable	grandeur f de rétroaction	Rückführgröße, Rückführvariable	переменная обратной связи	variabile di retroazione	variable f de realimentación, variable f realimentante	フィードバック変数 fidobakku-hensu	反馈变量 fankui bianliang

No.	English	French	German	Russian	Italian	Spanish	Japanese	Chinese
534	feedforward	action f prévisionnelle, anticipative, de tendance	Vorwärtszweig	прямая связь	azione diretta (in avanti)	anticipación f	フィードフォワード fido-fowādo	前馈 qiankui
535	feedforward compensation	compensation f à action prévisionnelle	Vorwärtskompensation	компенсация прямой связью	compensazione ad azione diretta	compensación f anticipativa	フィードフォワード補償 fidofowādo-hoshō	前馈补偿 qiankui buchang
536	feedforward control	régulation f à action prévisionnelle	Regelung mit Störgrößenaufschaltung	управление с прямой связью	controllo ad azione diretta	control m anticipativo	フィードフォワード制御 fidofowādo-seigyo	前馈控制 qiankui kongzhi
537	final controlled variable	grandeur f réglée finale	Endregelglied, Stellglied	предельная регулируемая величина, конечная регулируемая величина	variabile controllata finale	magnitud f regulada final	最終制御量 saishū-seigyoryō	末端被控变量 moduan beikong bianliang
538	final controlling drive	organe m de réglage final	Stellenantrieb	исполнительный механизм	azionamento terminale	actuador m final de control	操作部 sōsabu	末端控制驱动 moduan kongzhi qudong
539	final controlling element	organe m de réglage final	Stellglied	исполнительный механизм	elemento terminale di controllo	elemento m final de control		末端控制环节 moduan kongzhi huanjie
540	final value	valeur f réglée asymptotique	Endwert	установившееся значение	valore finale	valor m final	最終値 saishū-chi	终值 zhongzhi
541	final value theorem	théorème m de la valeur finale	Enwertsatz	теорема установившегося значения	teorema del valore finale	teorema * del valor final	最終値定理 saishūchi-teiri	终值定理 zhongzhi dingli
542	finite automatic machine	machine f à automates finis	endlicher Rechenautomat	конечный автомат	macchina automatica a stati finiti	máquina f automática finita		有限自动机器 youxian zidong jiqi
543	finite automaton	automate m fini	endlicher Automat	конечный автомат	automa a stati finiti	autómata m finito	有限オートマトン yūgen-ōtomaton	有限自动机 youxian zidongji
544	first-order system	système m du premier ordre	System mit Verzögerung erster Ordnung	система первого порядка	sistema del prim'ordine	sistema m de primer orden	1次系 ichiji-kei	一阶系统 yijie xitong
545	fixed command control (closed loop)	régulation f de maintien	Festwertregelung (Regelkreis)	стабилизирующее управление (замкнутое)	controllo a comando fissato (ad anello chiuso)	regulación f	定値制御	固定指令控制（闭环）guding zhiling kongzhi (bihuan)
546	flapper valve	vanne f à palettes	Klappenventil	клапанный вентиль, клапан	valvola a cerniera	paleta f	フラッパ furappa	舌形阀，挡板阀 shexingfa, dangbanfa

No.	English	French	German	Russian	Italian	Spanish	Japanese	Chinese
547	flat-card potentiometer	potentiomètre *m* bobiné à plat	Flachpotentiometer	плоский потенциометр	potenziometro planare	potenciómetro *m* plano	平たんカードポテンショメータ heitan-kādo-potensho-mēta	平片式电位计 pingpianshi dianweiji
548	flat-card resolver	résolveur *m* bobiné à plat	Flachresolver, Flach-Funktionsgeber	плоский решающий потенциометр	trasduttore planare di posizione angolare	trigonómetro *m*, potenciométrico plano	平たんカードレゾルバ heitan-kādo-rezoruba	平片式解算器 pingpianshi jiesuanqi
549	flip-flop	bascule *f* (bistable)	Flipflop, bistabiler Multivibrator	триггер	multivibratore bistabile (flip-flop)	báscula, flip-flop *m*	フリップフロップ furippu-furoppu	触发器，触发电路 chufaqi, chufa dianlu
550	floating action	action *f* flottante	Stelleingriff mit regeldifferenzabhängiger Stellgeschwindigkeit	астатическое воздействие	azione flottante	acción *f* flotante	積分動作 sekibun-dōsa	浮动作用 fudong zuoyong
551	floating control	commande *f* à action flottante	Regelung mit regeldifferenzabhängiger Stellgeschwindigkeit	астатическое управление	controllo flottante	control *m* flotante	積分制御 sekibun-seigyo	浮动控制 fudong kongzhi
552	floating controller	régulateur *m* flottant	Regler mit regeldifferenzabhängiger Stellgeschwindigkeit	астатический регулятор	controllore flottante	controlador *m* flotante	浮動調速器 fudō-seigyo-sōchi	浮动控制器 fudong kongzhiqi
553	floating lever	levier *m* à position flottante	Differentialhebel	маятник	leva flottante	palanca flotante		浮动杆 fudonggan
554	flow chart	graphe *m* de fluence	Flußdiagramm, Signalflußbild, Ablaufschema	поточный граф	schema di flusso	diagrama *m* de flujos, ordinograma *m*	流れ線図 nagare-senzu	流程图，程序框图 liuchengtu, chengxu kuangtu
555	flow control	régulation *f* de débit	Flußregelung, Ablaufregelung	управление потоком	controllo del flusso	control *m* de flujo	流量制御 ryūryō-seigyo	流量控制 liuliang kongzhi
556	flow diagram	diagramme *m* de fluence	Flußdiagramm, Fließbild, Signalflußdiagramm	поточная диаграмма	diagramma di flusso	diagrama *m* de flujo, ordinograma *m*	流れ図 nagarezu	程序框图，流程图 chengxu kuangtu, liuchengtu
557	fluidic device	circuit *m* fluidique	Fluidikelement, pneumatisches Logikelement	гидравлическое устройство	dispositivo fluidico	dispositivo *m* fluídico	流体論理素子 ryūtai-ronri-sōshi	流体装置，流体器件 liuti zhuangzhi, liuti qijian
558	flywheel	volant *m*	Schwungrad, Schwungscheibe	маховик	volano	rueda flotante	はずみ車 hazumi-guruma	飞轮 feilun
559	focus	foyer *m*	Brennpunkt	фокус	fuoco	foco *m*	焦点 shōten	焦点 jiaodian

No.	English	French	German	Russian	Italian	Spanish	Japanese	Chinese
560	follow-up control	régulation f de correspondance	Folgeregelung, Nachlaufregelung, Nachführregelung	управление с жёсткой обратной связью, следящее регулирование	controllo a seguire	control m de seguimiento (en anillo cerrado), servocontrol m	追從制御 tsuijū-seigyo	隨動控制 suidong kongzhi
561	force balance	équilibre m des forces	Kraftvergleich	равновесие сил	bilanciamento di forza	equilibrio m de fuerza	力平衡 chikaraheikō	力平衡 lipingheng
562	forced oscillation	oscillation f forcée	erzwungene Schwingung	вынужденные колебания	oscillazione forzata	oscilación f forzada	強制振動 kyōsei-shindō	強迫振蕩 qiangpo zhendang
563	formator	formateur m	Formierer	формирователь	formatore	formateador m	整書形 seishokei	格式器, 格式化程序 geshiqi, geshihua chengxu
564	forward channel	chaîne f d'action	Vorwärtskanal, Hauptkanal	канал прямой связи	linea di andata	camino m de acción, camino m directo	フォワードチャネル fowādo-chaneru	前饋信道，前饋通道 qiankui xindao, qiankui tongdao
565	forward controlling element	circuit m de régulation à action directe	Regelglieder, Steuerglieder (im Vorwärtszweig)	управляющий элемент в прямой связи	elemento di controllo ad azione diretta	elemento m de control directo	前向き調節要素 maemuki-chōsetsu-yōso	前饋控制環節 qiankui kongzhi huanjie
566	forward element	organe m d'action directe	Bauelement, Bauteil, Glied (im Vorwärtszweig)	элемент прямой связи	elemento ad azione diretta (in avanti)	elemento m de acción	前向き要素 maemuki-yōso	前饋環節，前饋元件 qiankui huanjie, qiankui yuanjian
567	forward path	chaîne f d'action	Vorwärtszweig, Vorwärtspfad	канал прямой связи	percorso diretto in avanti	camino m de acción, camino m directo	フォワード経路 fowādo-keiro	正向通路 zhengxiang tonglu
568	forward signal	signal m d'action	Signal im Vorwärtszweig, Vorwärtssignal	сигнал прямой связи	segnale diretto in avanti	señal f de acción	前向き信号 maemuki-shingō	前饋信号 qiankui xinhao
569	Fourier analysis	analyse f harmonique, analyse f de Fourier	Fourieranalyse	анализ Фурье, спектральный анализ	analisi di Fourier	analisis * de Fourier	フーリエ解析 furie-kaiseki	傅立叶分析 fuliye fenxi
570	Fourier expansion	développement m en série de Fourier	Fourier-Reihenentwicklung	разложение в ряд Фурье	sviluppo di Fourier	desarrollo m en serie de Fourier	フーリエ展開 furie-tenkai	傅立叶级数展开 fuliye jishu zhankaishi
571	Fourier integral	intégrale f de Fourier	Fourier-Integral	интеграл Фурье	integrale di Fourier	integral * de Fourier	フーリエ積分 furie-sekibun	傅立叶积分 fuliye jifen
572	Fourier series	série f de Fourier	Fourier-Reihe	ряд Фурье	serie di Fourier	series * de Fourier	フーリエ級数 furie-kyūsū	傅立叶级数 fuliye jishu

	English	French	German	Russian	Italian	Spanish	Japanese	Chinese
573	Fourier transform	transformée f de Fourier	Fourier-Transformation	преобразование Фурье	trasformata di Fourier	transformada f de Fourier	フーリエ変換 fūrie-henkan	傅立叶变换 fùlìyè biànhuàn
574	free motion	mouvement m libre	freie Bewegung	свободное движение	moto libero	movimiento m libre	自由運動 jiyū-undō	自由运动 zìyóu yùndòng
575	free oscillation	oscillation f libre	freie Schwingung	свободные колебания	oscillazione libera	oscilación f libre	自由振動 jiyū-shindō	自由振荡 zìyóu zhèndàng
576	frequency	fréquence f	Frequenz	частота	frequenza (pulsazione)	frecuencia f	周波数 shūhasū	频率 pínlǜ
577	frequency divider	diviseur m de fréquence	Frequenzteiler	делитель частоты	divisore di frequenza	divisor * de frecuencia	分周器 bunshuki	分频器 fēnpínqì
578	frequency domain	domaine m harmonique	Frequenzbereich	частотная область	dominio della frequenza	campo m frecuencial	周波数領域 shūhasū-ryo-iki	频率域 pínlǜyù
579	frequency response	réponse f harmonique	Frequenzgang, Frequenzverhalten	частотная характеристика	risposta in frequenza	respuesta f frecuencial, respuesta f armónica	周波数応答 shūhasū-ōtō	频率响应 pínlǜ xiāngyìng
580	frequency response characteristic	caractéristique f de la réponse harmonique	Frequenzgangkennlinie, Frequenzkennlinien	частотная характеристика	caratteristica della risposta in frequenza	característica f de respuesta armónica, característica f de respuesta frecuencial	周波数応答特性 shūhasū-ōtō-tokusei	频率响应特性 pínlǜ xiāngyìng tèxìng
581	frequency spectrum	spectre m de fréquence	Frequenzspektrum	частотный спектр	spettro di frequenza	espectro m de frecuencias	周波数スペクトル shūhasū-supekutoru	频谱 pínpǔ
582	friction	frottement m	Reibung	трение	attrito	rozamiento	摩擦 masatsu	摩擦力 mócalì
583	function (v)	fonctionner (v)	fungieren, tätig sein, funktionieren, wirksam sein	функционировать	funzionare	funcionar	作動する sadō-suru	起……功能作用 (动词) qǐ……gōngnéng zuòyòng (dòngcí)
584	function approximation	approximation f de fonction	Funktionsapproximation (od. Näherung)	аппроксимация функции, приближение функции	approssimazione di funzioni	approximación f de funciones	関数近似 kansū-kinji	函数通近 hánshu bījìn
585	function criterion	critère m de fonction	Funktionskriterium	функциональный критерий	cifra di merito	criterio * función		函数准则 hánshu zhǔnzé

	English	French	German	Russian	Italian	Spanish	Japanese	Chinese
586	function generator	générateur *m* de fonctions	Funktionsgenerator	функциональный генератор, функциональный преобразователь	generatore di funzione	generador *m* de función	関数発生器 kansū-hasseiki	函数发生器 hanshu fashengqi
587	functional block	bloc *m* fonctionnel	Funktionsblock	функциональный блок	blocco funzionale	bloque *m* funcional	機能ブロック kinō-burokku	功能块 gongneng kuai
588	functional chain	chaîne *f* fonctionnelle	Wirkungskette	функциональная цепь	catena funzionale	anillo *m* funcional	機能連鎖 kinō-rensa	功能链 gongneng lian
589	functional unit	élément *m* fonctionnel	Funktionseinheit	функциональный блок	unità funzionale	unidad *f* funcional	機能ユニット kinō-yunitto	功能单元 gongneng danyuan
590	fundamental input-connection matrix	matrice *f* à connections d'entrées fondamentales	Eingangs (verbindungs)matrix	фундаментальная матрица входов	matrice fondamentale di connessione agli ingressi	matriz de conexión de entrada fundamental		基本输入连接矩阵 jiben shuru lianjie juzhen
591	fundamental interconnection vector	vecteur *m* d'interconnections fondamentales	fundamentaler (od. grundlegender) Kopplungsvektor	фундаментальный вектор взаимных соединений	vettore d'interconnessione fondamentale	vector de interconexión fundamental		基本互联向量 jiben hulian xiangliang
592	fundamental matrix	matrice fondamentale	Fundamentalmatrix	фундаментальная матрица	matrice fondamentale	matriz *f* fundamental	基本帰行列 kihonkai-gyōretsu	基本矩阵 jiben juzhen
593	fuzziness	flou *m*	Unschärfe	нечёткость, размытость	indeterminazione	borrosidad *f*	あいまいさ aimaisa	模糊度 mohudu
594	fuzzy logic	logique *f* floue	Fuzzy-Logik, unscharfe Logik	размытая (=нечёткая) логика	logica 'fuzzy' (logica "sfumata")	lógica *f* difusa, lógica *f* borrosa	ファジー論理 faji-ronri	模糊逻辑 mohu luoji
595	fuzzy set	ensemble *m* flou	Fuzzy-Menge	размытое множество	insieme 'fuzzy' (insieme "sfumato")	conjunto * borroso	ファジー集合 faji-shūgo	模糊集 mohuji
596	fuzzy system	système *m* flou	Fuzzy-System	размытая система	sistema 'fuzzy' (sistema "sfumato")	sistema *m* difuso, sistema *m* borroso	ファジーシステム faji-shisutemu	模糊系统 mohu xitong
597	gain	gain *m*	Verstärkung, Gewinn, Zunahme	коэффициент усиления, усиление	guadagno	ganancia *f*	ゲイン gein	增益 zengyi
598	gain characteristic	caractéristique *f* de gain	Verstärkungscharakteristik	коэффициент усиления	diagramma dei moduli	característica de ganancia	ゲイン特性 gein-tokusei	增益特性 zengyi texing

No.	English	Deutsch	Français	Русский	Italiano	Español	日本語	中文
599	gain cross-over frequency	Durchtrittsfrequenz (Amplitudenverlauf)	fréquence f de coupure	частота разделения по коэффициенту усиления	frequenza (pulsazione) di taglio (o d'intersezione)	frecuencia f de la ganancia de corte	ゲイン交点周波数 gein-kōten-shūhasū	增益穿越频率 zengyi chuanyue pinlü
600	gain margin	Amplitudenrand	marge f de gain	запас устойчивости	margine di guadagno	margen m de gananacia	ゲイン余有 gein-yoyū	增益裕度 zengyi yudu
601	gate element	Tor, Gitter, Torschaltung, Gatter	porte f (logique)	вентиль	porta	(elemento) puerta f	ゲート要素 gēto-yōso	门元件 men yuanjian
602	Gaussian distribution	Gaußverteilung	répartition f normale	нормальное распределение, гауссовское распределение	distribuzione gaussiana	distribución f de gauss, distribución f normal	ガウス分布 gausu-bunpu	高斯分布, 正态分布 gaosi fenbu, zhengze fenbu
603	Gaussian plane	Gaußsche Zahlenebene	plan m de Gauss	комплексная плоскость	piano di Gauss	plano m complejo	ガウス平面 gausu-heimen	高斯平面 gaosi pingmian
604	gimbal	Kardanring (Kreisel), kardanische Aufhängung	anneau m de Cardan	карданный подвес, универсальный шарнир	sospensione cardanica	suspensión f de Cardán	ジンバル jinbaru	框架 kuangjia
605	gimbal axis	Rahmenachse (Kreisel)	axe m de Cardan	ось кардана	asse di una sospensione cardanica	eje m de la suspensión (de Cardán)	ジンバル角 jinbaru-kaku	框架轴 kuangjia zhou
606	govern (v)	regeln (Drehzahl), leiten, lenken	régler (v)	управлять	governare	regular	調速する chōsoku-suru	调节, 控制 (动词) tiaojie, kongzhi (dongci)
607	governor	Drehzahlregler, (Fliehkraft)regler	régulateur m de vitesse	регулятор скорости	controllore	regulador m de velocidad	調速器 chōsokuki gabana	调速器, 调节器 tiaosuqi, tiaojieqi
608	graded potentiometer	Stufenpotentiometer	potentiomètre m non-linéaire	нелинейный потенциометр	potenziometro graduato	potenciómetro m no lineal	分段電位計	分段电位计 fenduan dianweiji
609	gradient method	Gradientenmethode	méthode f du gradient	градиентный метод	metodo del gradiente	método m del gradiente	勾配法 keisha-hō	梯度法 tidufa
610	graphic display	graphische Anzeige, Grafiksichtgerät	affichage m graphique	графический дисплей	visualizzatore grafico	pantalla f gráfica	グラフィックディスプレイ zukei-hyōji-sōchi	图形显示 tuxing xianshi
611	graphic printer	Grafikdrucker	imprimante f graphique	графический принтер, графическое печатающее устройство	stampante grafica	impresora f gráfica	グラフィックプリンタ zukei-insatsu-sōchi	图形打印机 tuxing dayinji

	English	French	German	Russian	Italian	Spanish	Japanese	Chinese
612	Germ's function	fonction f de Green	Greensche Funktion	функция Грина	funzione di Green	funcíon de Green	グリーン関数 gurīn-kansū	格林函数 gelin hanshu
613	guidance system	système m de conduite	Leitsystem	система наведения	sistema di guida	sistema m de guiado	誘導システム yūdō-shisutemu	制导系统 zhidao xitong
614	gyro	gyroscope m	Kreisel (-gerät)	гироскоп	giroscopio	giroscopio m	ジャイロ jairo	陀螺 tuoluo
615	gyroscope	gyroscope m	Kreisel	гироскоп	giroscopio	giroscopio m	ジャイロスコープ jairo-sukōpu	陀螺仪 tuoluo yi
616	hard manual control	régulation f manuelle	manueller Eingriff unter erschwerten Bedingungen	жёстко-ручное управление	controllo manuale diretto	control manual estricto		手动硬控制 shoudong yingkongzhi
617	harmonic function	fonction f harmonique	harmonische Funktion	гармоническая функция	funzione armonica	función f armónica	調和関数 chōwa-kansū	谐波函数 xiebo hanshu
618	harmonic response	réponse f harmonique	harmonische Systemantwort, Frequenzgang	частотная характеристика	risposta armonica	respuesta f armónica, respuesta f frecuencial	調和応答 chōwa-ōtō	谐波响应 xiebo xiangying
619	harmonic response characteristic	caractéristique f de réponse harmonique	Frequenzgang-charakteristik	частотная характеристика	caratteristica della risposta armonica	caracteristica f de respuesta armónica, caracteristica f de respuesta frecuencial	調和応答特性 chōwa-ōtō-tokusei	谐波响应特性 xiebo xiangying texing
620	hierarchical control	régulation f hiérarchisée	hierarchische Regelung	иерархическое управление	controllo gerarchico	control m jerárquico	階層制御 kaisō-seigyo	递阶控制 dijie kongzhi
621	hierarchical structure	structure f hiérarchisée	hierarchische Struktur, hierarchischer Aufbau	иерархическая структура	struttura gerarchica	estructura f jerárquica	階層構造 kaisō-kōzō	递阶结构 dijie jiegou
622	hierarchical system	système m hiérarchisé	hierarchisches System	иерархическая система	sistema gerarchico	sistema m jerárquico	階層システム kaisō-shisutemu	递阶系统 dijie xitong
623	high level language	langage m élaboré, de haut niveau	höhere Programmiersprache, Hochsprache	язык высокого уровня	linguaggio d'alto livello	lenguaje * de alto nivel	高レベル言語 kōreberu-gengo	高级语言 gaoji yuyan
624	high-pass filter	filtre m passe-haut	Hochpaßfilter	фильтр (пропускания) высоких частот	filtro passa alto	filtro * pasa-alto	高域フィルタ kō-iki-firuta	高通滤波器 gaotong lüboqi

No.	English	French	German	Russian	Italian	Spanish	Japanese	Chinese
625	hill climbing	recherche f de l'extremum	Gradientenmethode (steilster Aufstieg)	поиск экстремума	salita	Búsqueda f de extremos	山登り法 yamanobori-hō	爬山法 pashanfa
626	holding	maintien m	Haltevorgang	блокировка, удержание	tenuta	bloqueo, mantenimiento m, retención f	保持 hoji	保持 baochi
627	holding action	action f de maintien	Halteoperation	воздействие с запоминанием	azione di tenuta	acción f de bloqueo, acción f de mantenimiento, acción f de retención	保持動作 hoji-dōsa	保持作用 baochi zuoyong
628	holding element	organe m de maintien	Halteglied	фиксирующий элемент, запоминающий элемент	elemento di tenuta	(elemento) bloqueador m, (elemento) mantenedor m, (elemento) retenedor m bombeo m, penduleo m	保持要素 hoji-yōso	保持环节 baochi huanjie
629	hunting	pompage m	pendeln, schwingen, Drehzahlschwankungen	бросание	caccia, ricerca		乱調 ranchō	振颤 zhenchan
630	Hurwitz polynomium	polynôme m hurwitzien	Hurwitzpolynom	полином Гурвица	polinomio di Hurwitz	polinomio m de Hurwitz	フルヴィッツ多項式 Hurwitz-takōshiki	赫尔维茨多项式 heerweici duoxiangshi
631	Hurwitz's criterion	critère m de Hurwitz	Hurwitzkriterium	критерий Гурвица	criterio di Hurwitz	criterio m de Hurwitz	フルヴィッツの（安定）基準 Hurwitz-no-(antei)kijun	赫尔维茨准则 heerweici zhunze
632	hybrid computer	calculateur m hybride, calculatrice f hybride	Hybridrechner	гибридная вычислительная машина	calcolatore ibrido	computador m hibrido	ハイブリッド計算機 haiburiddo-keisanki	混合计算机 hunhe jisuanji
633	hydraulic accumulator	accumulateur m hydraulique	Hydraulikspeicher, hydraulischer Speicher	гидравлический аккумулятор	accumulatore idraulico	acumulador m hidráulico	流体アキュムレータ ryutai-akyumurēta	减压储能器 yeya chu'nengqi
634	hydraulic actuator	servomoteur m hydraulique	Hydraulikantrieb	гидропривод	attuatore idraulico	accionador m hidráulico	流体アクチュエータ ryutai-akuchuēta	液动执行机构 yedong zhixing jigou
635	hydraulic amplifier	amplificateur m hydraulique	Hydraulikverstärker	гидроусилитель	amplificatore idraulico	amplificador m hidráulico	流体増幅器 ryutai-zōfukuki	液压放大器 yeya fangdaqi
636	hydraulic motor	moteur m hydraulique	Hydraulikmotor	гидромотор	motore idraulico	motor m hidráulico	油圧モータ yuatsu-mōta	液压马达 yeya mada
637	hydraulic relay	relais m hydraulique	Hydraulikrelais	гидрореле	'relay' (o commutatore) idraulico	relé m hidráulico	流体リレー ryutai-rirē	液压继电器 yeya jidianqi

№	English	French	German	Russian	Italian	Spanish	Japanese	Chinese
638	hyperstability	hyperstabilité f	Hyperstabilität	гиперустройчивость	iperstabilità	hiperestabilidad f	超安定 chōantei	超稳定性 chaowendingxing
639	hysteresis	hystérésis f	Hysterese	гистерезис	isteresi	histéresis f	ヒステリシス hisuterishisu	磁滞残余；寿后作用 cizhi xianxiang, zhihou zuoyong
640	hysteresis error	erreur f de réversibilité	Hysteresefehler	ошибка гистерезиса	errore d'isteresi	error m de histéresis	ヒステリシス誤差 hisuterishisu-gosa	滞环误差 zhihuan wucha
641	hysteresis motor	moteur m à hystérésis	Hysteresemotor	гистерезисный двигатель	motore a isteresi	motor m de histeresis	ヒステリシスモータ hisuterishisu-mōta	磁滞马达 cizhi mada
642	ideal final value	valeur f désirée	idealer Endwert	заданное конечное значение	valore finale ideale	valor m final ideal	理想最終値 risō-saishū-chi	理想终值 lixiang zhongzhi
643	ideal value	valeur f prescrite	idealer Wert	заданное значение	valore ideale	valor m ideal	目標値 mokuhyō-chi	理想值 lixiangzhi
644	identifiability	identifiabilité f	Identifizierbarkeit	идентифицируемость	identificabilità	identificabilidad f	可同定性 ka-dōteisei	可辨识性 kebianshixing
645	identification	identification f	Identifizierung	идентификация	identificazione	identificación f	同定 dōtei	辨识 bianshi
646	image processing	traitement m d'images	Bildverarbeitung	обработка изображений	elaborazione d'immagini	tratamiento m de imágenes	画像処理 gazō-shori	图象处理 tuxiang chuli
647	image converter	convertisseur m d'images	Bildwandler	преобразователь изображений	convertitore d'immagini	convertidor • de Imagen	イメージ変換器 imēji-henkanki	图象变换器 tuxiang bianhuanqi
648	image intensifier	amplificateur m d'images	Bildverstärker	интенсификатор изображений	intensificatore d'immagini	intensificador • de Imagen	イメージ増幅器 imēji-zōfukuki	图象强化器 tuxiang qianghuaqi
649	imaginary axis	axe m imaginaire	imaginäre Achse	мнимая ось	asse immaginario	eje m imaginario	虚軸 kyojiku	虚轴 xuzhou
650	impedance	impédance f	Impedanz, Scheinwiderstand	полное сопротивление	impedenza	impedancia f	インピーダンス inpīdansu	阻抗 zukang

	English	French	German	Russian	Italian	Spanish	Japanese	Chinese
651	implement (v)	mettre en oeuvre (v), réaliser (v), exécuter (v)	realisieren., verwirklichen, implementieren	применять, внедрять, осуществлять	implementare, realizzare	realizar, llevar-a-cabo	実行する jikkō-suru	实现，执行（动词）shixian, zhixing(dongci)
652	impulse	impulsion f	Impuls, Stoß	импульс	impulso	impulso m	インパルス inparusu	脉冲 chongji liang
653	impulse function	fonction f impulsionnelle	Impulsfunktion, Stoßfunktion	импульсная функция	funzione impulsiva	función f impulso	インパルス関数 inparusu-kansū	脉冲函数 chongji hanshu
654	impulse response	réponse f impulsionnelle	Impulsantwort	импульсная характеристика	risposta all'impulso	respuesta f impulsional	インパルス応答 inparusu-ōtō	脉冲响应 chongji xiangying
655	impulse signal	signal m impulsionnel	Impulssignal	импульсный сигнал	impulso	señal f de impulso	インパルス信号 inparusu-shingō	脉冲信号 chongji xinhao
656	in-line	en ligne	innerhalb des Systems, gleichlaufend, ohne Verzögerung	пущенный в эксплуатацию	in linea	en linea	埋め込み uzumekomi	线内 xiannei
657	increasing oscillation	oscillation f croissante	aufklingende Schwingung	расходящиеся колебания	oscillazione crescente	oscilación f creciente	増大振動 zōdai-shindō	增幅振荡 zengfu zhendang
658	incremental	incrémental	inkremental, zunehmend, anwachsend	постепенно нарастающий, инкрементный	incrementale	incremental	増分 zōbun	增量的 zengliangde
659	independent variable	grandeur f indépendante	unabhängige Variable	независимая переменная	variabile indipendente	variable f independiente	独立変数 dokuritsu-hensū	自变量 zibianliang
660	indicated angle	angle m affiché	Winkelanzeige	угол индикации	angolo indicato	ángulo m indicado	指示角度 shiji-kakudo	指示角 zhishi jiao
661	indicial response	réponse f indicielle	Einheitssprungantwort	переходная характеристика	risposta indiciale	respuesta f a un escalón unitario, respuesta f unitaria	インディシャル応答 ステップ応答 indisyaru-ōtō suteppu-ōtō	特征响应 zhishu xiangying
662	indirectly controlled system	système m de régulation indirecte	indirekt geregeltes System	косвенно управляемая система	sistema controllato indirettamente	sistema controlado indirectamente		间接被控系统 jianjie beikong xitong
663	indirectly controlled variable	grandeur f réglée de manière indirecte	indirekte Regelgröße	косвенно управляемая переменная	variabile controllata indirettamente	variable f indirectamente controlada	間接制御量 kansetsu-seigyo-ryō	间接被控变量 jianjie beikong bianliang

№	English	French	German	Russian	Italian	Spanish	Japanese	Chinese
664	induction motor	moteur *m* à induction	Induktionsmotor	индукционный мотор	motore a induzione	motor *m* de inducción	誘導電動機 yūdō-dendōki	感应电动机 ganying diandongji
665	inductive pick-off	capteur *m* inductif	induktive Abnahme	индуктивный датчик	captatore induttivo	selección inductiva	誘導検出 yūdō-kenshutsu	感应式传感器 ganyingshi chuanganqi
666	inductor	inducteur *m*	Induktor, Drosselspule, Läufer	катушка индуктивности	induttore	inductor *m*	誘導子 yūdōshi	感应器 ganyingqi
667	inertia	inertie *f*	Trägheit, Beharrungsvermögen	инерция	inerzia	inercia *f*	慣性 kansei	惯性，惰量 guanxing, guanliang
668	infinite series	série *f* infinie	unendliche Reihe	бесконечный ряд	serie infinita	serie * infinita	無限級数 mugen-kyūsū	无穷级数 wuqiong jishu
669	information analysis	analyse *f* de l'information	Informationsanalyse	анализа информации	analisi dell'informazione	análisis * de la información	情報解析 jōhō-kaiseki	信息分析 xinxi fenxi
670	information parameter	paramètre *m* d'information	Informationsparameter	информационный параметр	parametro d'informazione	parámetro *m* de información	情報パラメータ jōhō-paraméta	信息参量 xinxi canliang
671	information pattern	représentation *f* de l'information	Informationsmuster	информационный образ	struttura informativa	patrón *m* de información	情報パターン jōhō-patān	信息模式 xinxi moshi
672	information retrieval	rétablissement *m* de l'information	Informationsschließung, Wiedergewinnung von Nachrichten	поиск информации	reperimento di informazioni	recuperación *f* de la información	情報検索 jōhō-kensaku	信息检索 xinxi jiansuo
673	information technology (IT)	technologie *f* de l'information	Informationstechnik	информатика, информационная технология	tecnologia dell'informazione (IT)	tecnología * de la información	情報技術 jōhō-gijyutu	信息技术 xinxi jishu
674	information theory	théorie *f*	Informationstheorie	теория информации	teoria dell'informazione	teoría * de la información	情報理論 jōhō-riron	信息论 xinxi lilun
675	inherent feedback	auto-réaction *f*, réaction *f* propre	innere Rückführung	собственная (внутренняя) обратная связь	retroazione intrinseca	realimentación *f* intrínseca	固有のフィードバック koyū-no-fīdobakku	固有反馈 guyou fankui
676	inherent proportional band	bande étendue *f* proportionnelle	inhärenter Proportionalbereich	собственная зона пропорционального регулирования	banda proporzionale intrinseca	banda proporcional inherente	固有の比例帯 koyū-no-hireitai	固有比例带 goyou bilidai

No.	English	Français	Deutsch	Русский	Italiano	Español	日本語	中文
677	inherent regulation	auto-régulation f	17F Spannungsänderung (bei konst. Drehzahl), Drehzahländerung (bei konst. Spannung u. 678requenz)	саморегулирование	regolazione intrinseca	autorregulación, regulación f intrínseca		固有调节 guyou tiaojie
678	inherent stability	stabilité f propre	Eigenstabilität, inhärente Stabilität	собственная устойчивость	stabilità intrinseca	estabilidad f intrínseca, estabilidad f propia	固有の安定性 koyū-no-anteisei	固有稳定性 guyou wendingxing
679	initial state	état m initial	Anfangszustand	исходное (=начальное) состояние	stato iniziale	estado m inicial	初期状態 shoki-jōtai	初始状态 chushi zhuangtai
680	inner gimbal	Cardan m intérieur	innerer Kardanrahmen	внутренний карданный подвес	sospensione cardanica interna	suspensión f cardánica interior	内部ジンバル naibu-jinbaru	内框 neikuang
681	inner loop	boucle f secondaire	innerer Kreis, innere (Programm)schleife	внутренний контур	anello interno	anillo m interior, anillo m secundario	内部ループ naibu-rūpu	内环 neihuan
682	input	grandeur f d'entrée	Eingangsgröße	вход	ingresso	entrada f	入力 nyūryoku	输入 shuru
683	input-centralized system	système m à accès centralisés	eingangszentralis-iertes System	система с централизованным входом	sistema a ingresso centralizzato	sistema m de entradas centralizadas		集中输入式系统 jizhong shurushi xitong
684	input-decentralized system	système m à accès décentralisés	eingangsdezentralisiertes System	система с децентрализованным входом	sistema a ingresso decentralizzato	sistema m de entradas descentralizadas		分散输入式系统 fensan shurushi xitong
685	input element	organe m d'entrée	Eingabeglied	входный элемент	elemento d'ingresso	elemento m de entrada	入力要素 nyūryoku-yōso	输入环节、输入元件 shuru huanjie, shuru yuanjian
686	input equipment	appareillage m d'entrée	Eingangsgerät, Eingangsbetriebsmittel	входное устройство	apparecchiatura d'ingresso	equipo m de entrada	入力装置 nyūryoku-sōchi	输入装置 shuru zhuangzhi
687	input matrix	matrice f d'entrée	Eingangsmatrix	матрица входа	matrice d'ingresso	matriz f de entrada	入力行列 nyūryoku-gyōretsu	输入矩阵 shuru juzhen
688	input power winding	enroulement m d'alimentation	Eingangleistungs-wicklung	питающая обмотка, входная силовая обмотка	avvolgimento di potenza d'ingresso	arrollamiento m de alimentación, devanado m de alimentación	入力電力巻線 nyūryoku-dennyoku-makisen	输入电源绕组 shuru dianyuan raozu
689	input signal	signal m d'entrée	Eingangssignal	входной сигнал, входное воздействие	segnale d'ingresso	señal f de entrada	入力信号 nyūryoku-shingō	输入信号 shuru xinhao

No.	English	French	German	Russian	Italian	Spanish	Japanese	Chinese
690	input variable	variable f d'entrée	Eingangsgröße, Eingangsvariable	входная величина	variabile d'ingresso	magnitud f de entrada, variable f de entrada	入力変数 nyūryoku-hensū	输入变量 shuru bianliang
691	insensitive	insensible	unempfindlich	нечувствительный	insensitivo	insensible	鈍感な donkan-na	不灵敏的, 不敏感的 bulingminde, bumin gande
692	insensitivity	insensibilité f	Unempfindlichkeit, Empfindungslosigkeit	нечувствительность	insensitività	insensibilidad f	鈍感 donkan	不灵敏性 bulingminxing
693	instability	instabilité	Instabilität	неустойчивость	instabilità	inestabilidad	不安定性 fuanteisei	不稳定性 buwendingxing
694	instantaneous value	valeur f instantanée	Momentanwert	мгновенное значение	valore istantaneo	valor m instantáneo	瞬間値 shunkan-chi	瞬时值 shunshizhi
695	integer programing	programmation f en nombres entiers	ganzzahlige Programmierung	целочисленное программирование	programmazione a numeri interi	programación * entera	整数計画法 seisū-keikakuhō	整数规划 zhengshu guihua
696	integral action	action f intégratrice, action f l	integrierendes Verhalten, Integralverhalten	интегрирующее воздействие	azione integrale	acción f integral, acción f integral	積分動作 sekibun-dōsa	积分作用 jifen zuoyong
697	integral action factor	coefficient m d'action par intégration	Integrierbeiwert, integraler Übertragungsfaktor	коэффициент интегрирующего воздействия, постоянная интегрирования	fattore (o coefficiente) dell'azione integrale	coeficiente m de acción integral	積分動作係数 sekibun-dōsa-keisū	积分作用因子 jifen zuoyong yinzi
698	integral action rate	taux m d'action par intégration	Integrationsgeschwindigkeit	скорость интегрального воздействия	velocità dell'azione integrale	razón f de acción integral	積分動作速度 sekibun-dōsa-jikan	积分作用速率 jifen zuoyong sulü
699	integral action time	temps m de dosage d'intégration, de compensation	Integrierzeit	время интегрирования	tempo dell'azione integrale	tiempo m de acción integral	積分動作時間 sekibun-dōsa-jikan	积分作用时间 jifen zuoyong shijian
700	integral controller	régulateur m intégral	Integralregler, I-Regler	интегральный регулятор, И-регулятор	controllore integrale	regulador m de acción integral	積分形調節器 sekibun-gata-chōsetsuki	积分控制器 jifen kongzhiqi
701	integral element	organe m d'action par intégration, d'action flottante	integrierendes Element, I-System	интегрирующий элемент	elemento integrale	elemento m de acción integral	積分要素 sekibun-yōso	积分环节, 积分元件 jifen huanjie, jifen yuanjian
702	integral equation	équation f intégrale	Integralgleichung	интегральное уравнение	equazione integrale	ecuación * integral	積分方程式 sekibun-hōteishiki	积分方程 jifen fangcheng

#	English	French	German	Russian	Italian	Spanish	Japanese	Chinese
703	integrating action	action f intégratrice	integrales Verhalten	интегрирующее воздействие	integrazione	acción f integradora	積分動作 sekibun-dōsa	积分作用 jifen zuoyong
704	integrating amplifier	amplificateur m intégrateur	Integrationsverstärker, I-Verstärker	интегрирующий усилитель	amplificatore d'integrazione	amplificador m integrador	積分増幅器 sekibun-zōfukuki	积分放大器 jifen fangdaqi
705	integrating element	organe m d'intégration	Bauelement mit Integralverhalten	интегрирующий элемент	elemento integrante	(elemento) integrador m	積分要素 sekibun-yōso	积分环节，积分元件 jifen huanjie, jifen yuanjian
706	integrating gyro	gyroscope m intégrateur	Integrationskreisel	интегрирующий гироскоп	giroscopio integratore	giroscopio m integrador	積分ジャイロ sekibun-jairo	积分陀螺 jifen tuoluo
707	integrator	intégrateur m	Integrator	интегратор, интегрирующий элемент	integratore	integrador m	積分器 sekibun-ki	积分器 jifenqi
708	integro-differential equation	équation f intégro-différentielle	Integro-Differentialgleichung	интегро-дифференциальное уравнение	equazione integro-differenziale	ecuación * integro-diferencial	積分-微分方程式 sekibun-bibun-hōteishiki	积分-微分方程 jifen-weifen fangcheng
709	intelligent knowledge-based system (IKBS)	système m intelligent à base de connaissances	intelligentes wissensbasiertes System	интеллектуальная система базирующаяся на знаниях	sistema intelligente basato sulla conoscenza	sistema inteligente dasado enel conocimiento	知的知識ベースシステム chiteki-chishikibēsu-shisutemu	智能知识库系统 zhineng zhishiji xitong
710	interaction	interaction f	Wechselwirkung, gegenseitige Beeinflussung, Interaktion	интеракция, взаимодействие	interazione	interacción f	相互干渉 sōgo-kanshō	互联 hulian
711	interactive program	programme m interactif	interaktives Programm	интерактивная программа	programma interattivo	programa * interactivo	会話型プログラム	交互程序 jiaohu chengxu
712	interconnected network	réseau m interconnecté	vermaschtes (od. verkoppeltes) Netzwerk, Netzverbund	связанная сеть	rete interconnessa	red f interconectada	結合回路網 ketsugō-kairo-mō	互联网络 hulian wangluo
713	interconnected subsystems	sous-systèmes m.pl interconnectés	vermaschtes Teilsystem, Teil-Verbundsystem	связанные подсистемы	sottosistema interconnesso	subsistemas m interconectados	結合サブシステム ketsugō-sabushisutemu	互联子系统 hulian xitong
714	interconnection matrix	matrice f d'interconnexion	Kopplungsmatrix	матрица связи	matrice d'interconnessione	matriz f de interconexión	相互結合行列 sōgoketsugō-gyōretsu	互联矩阵 hulian juzhen
715	interface	interface f	Interface, Schnittstelle, Grenzschicht	интерфейс, сопряжение	interfaccia	interfase f	インターフェイス intāfeisu	界面，接口 jiemian, jiekou

	English	French	German	Russian	Italian	Spanish	Japanese	Chinese
716	interface system	système m d'interfaçage	Schnittstellensystem, Prozeßperipherie	система сопряжения	sistema d'interfaccia	sistema de interfaz	インターフェイスシステム intāfeisu-shisutemu	界面系统, 接口系统 jiemian xitong, jiekou xitong
717	interface unit	élément m d'interfaçage	Schnittstelleneinheit	устройство сопряжения	unità d'interfaccia	unidad de interfaz	インターフェイスユニット intāfeisu-yunitto	界面单元, 接口单元 jiemian danyuan, jiekou danyuan
718	interlocking	verrouillage m mutuel	Verriegelungskreis, Schaltverriegelung, Sicherheitsschaltung	взаимоблокировка	interallacciamento, sincronizzazione	enclavamiento m mútuo	インターロック intārokku	联锁 liansuo
719	intermittent signal	signal m intermittent	intermittierendes (auch diskontinuierliches) Signal	прерывистый сигнал	segnale intermittente	señal f intermitente	間欠信号 kanketsu-shingō	间发信号 jianfa xinhao
720	interpreter	interpréteur m	Übersetzer, Dolmetscher	интерпретатор	interprete	interpretador m	インタプリタ（通訳ルーチン）intāpurita (tūyaku-rūchin)	解释程序 jieshi chengxu
721	interrupt	interruption f	Unterbrechung	прерывание	interruzione	interrupción f	割り込み warikomi	中断 zhongduan
722	invariance	invariance f	Invarianz, Unveränderlichkeit	инвариантность	invarianza	invariancia *	不変性 fuhensei	不变性 bubianxing
723	inverse Nyquist array	réseau m de Nyquist inverse	inverses Nyquistschema	обратный ряд Найквиста	quadro di Nyquist inverso	lugar inverso de Nyquist	逆ナイキスト配列 gyaku-naikisuto-hairetsu	逆奈奎斯特阵列 ninaikuisite zhenlie
724	inverse system	système m inverse	inverses System	обратная система	sistema inverso	sistema m inverso	逆システム gyaku-shisutemu	逆系统 nixitong
725	inverse transfer function	transmittance f inverse	inverse (komplexe) Übertragungsfunktion	обратная передаточная функция	funzione di trasferimento inversa	función f de transferencia inversa, transmitancia f inversa	逆伝達関数 gyaku-dentatsu-kansū	逆传递函数 nichuandihanshu
726	inverse transfer locus	lieu m de transfert inverse	Ortskurve der inversen Übertragungsfunktion	обратная амплитудно-фазовая характеристика	diagramma di trasferimento inverso	lugar m (geométrico) de la función, de transferencia inversa	逆伝達軌跡 gyaku-dentatsu-kiseki	逆传递轨迹 nichuandiguiji
727	inverse transform	transformée f inverse	Rücktransformation, inverse Transformation	оригинал, обратное преобразование	antitrasformata	transformada f inversa	逆変換 gyaku-henkan	反变换 fanbianhuan
728	isolated network	réseau m indépendant	isoliertes Network	изолированная сеть	rete isolata	red f aislada	孤立回路網 koritsu-kairo-mō	孤立网络 guli wangluo

#	English	Français	Deutsch	Русский	Italiano	Español	日本語	中文
729	iterative control algorithm	algorithme *m* de réglage itératif	iterativer Regel- (od. Steuer-)Algorithmus	итеративный алгоритм управления	algoritmo di controllo iterativo	algoritmo *m* iterativo de control	繰り返し制御アルゴリズム kurikaeshi-seigyo-arugorizumu	迭代控制算法 diedai kongzhi suanfa
730	Jacobi matrix	matrice *f* jacobienne	Jacobi-Matrix	матрица Якоби	matrice di Jacobi (o jacobiana)	matriz * de Jacobi	ヤコービ行列 yakōbi-gyōretsu	雅可比矩阵 yakebi juzhen
731	jet-pipe	gicleur *m*	Strahlrohr	струйная трубка, сопло	tubo di scarico	tobera *f*	噴射管 funshakan	射流管 sheliu guan
732	joint probability	densité *f* de probabilité	Verbundwahr-scheinlichkeit	совместная вероятность	probabilità congiunta	probabilidad * conjunta	共確率	联合概率 lianhe gailü
733	Jordan normal form	forme *f* normale de Jordan	Jordansche Normalform	жорданова нормальная форма, каноническая форма Жордана	forma normale di Jordan	forma *f* normal de Jordan	ジョルダンの標準形 Jordan-no-hyōjunkei	约当标准型 yuedang zhengguixing
734	Jordanian canonical form	forme *f* canonique de Jordan	Kanonische Jordanform	жорданова нормальная форма, каноническая форма Жордана	forma canonica di Jordan	forma *f* conónica de Jordan	ジョルダンの標準形 Jordan-no-hyōjunkei	约当型 yuedang biaozhunxing
735	Kalman filter	filtre *m* de Kalman	Kalman-Filter	фильтр Кальмана	filtro di Kalman	filtro *m* de Kalman	カルマンフィルタ Kalman-firuta	卡尔曼滤波器 kaerman liuboqi
736	keyboard	clavier *m*	Tastenfeld, Tastatur	клавиатура	tastiera	teclado *m*	キーボード（鍵盤）kībōdo(kenban)	键盘 jianpan
737	kinetic control system	système *m* de réglage cinématique	kinetisches Regelungssystem	кинематическая система управления	sistema di controllo cinetico	sistema *m* de control cinemático	力学的制御系 rikigaku-teki-seigyokei	运动学控制系统 yundongxue kongzhi xitong
738	knowledge-based system (KBS)	système *m* fondé sur l'exploitation d'une base de connaissance	wissensbasiertes System	система базирующася на знаниях	sistema basato sulla conoscenza (KBS)	sistemas * basado en el conocimiento	知識ベースシステム chishikibēsu-shisutemu	知识库系统 zhishiji xitong
739	lag	retard *m*	Verzögerung	замедление, инертное замедление	ritardo	retardo *m*	おくれ okure	滞后 zhihou
740	lag element	organe *m* de retard	Verzögerungsglied	апериодический элемент	elemento di ritardo	elemento *m* de retardo	おくれ要素 okure-yōso	滞后环节, 滞后元件 zhihou huanjie, zhihou yuanjian
741	lag network	réseau *m* à retard	verzögerndes Netzwerk, Verzögerungsglieder	апериодическая сеть (цепь)	rete ritardatrice	red de retraso (retardo)	おくれ回路 okure-kairo	滞后网络 zhihou wangluo

	English	French	German	Russian	Italian	Spanish	Japanese	Chinese
742	Laplace transform	transformée f de Laplace	Laplace-Transformierung	преобразование Лапласа, изображение по Лапласу	trasformata di Laplace	transformada f de Laplace	ラプラス変換 Laplace-henkan	拉普拉斯变换 lapulasi bianhuan
743	Laplace transformation	transformation f de Laplace	Laplace-Transformation	преобразование Лапласа	trasformazione di Laplace	transformación f de Laplace	ラプラス変換 Laplace-henkan	拉普拉斯变换 lapulasi bianhuan
744	large-scale system	grand système m	Großsystem	большая система	sistema a grandi dimensioni	sistema m de gran escala	大規模系 (システム) daikibo-kei (-shisutemu)	大系统 daxitong
745	laser printer	imprimante f laser	Laserdrucker	лазерный принтер	stampante laser	impresora * láser	レーザプリンタ rēza-purinta	激光打印机 jiguang dayinji
746	lead network	réseau m d'avance de phase	voreilendes Netzwerk, Vorhaltglieder	цепь опережения	rete anticipatrice	red f de adelanto de fase, red f de avance de fase	位相すすみ回路 isō-susumi-kairo	超前网络 chaoqian wangluo
747	learning system	système m d'apprentissage	lernfähiges System	обучающая система	sistema ad apprendimento	sistema m de autoaprendizaje	学習システム gakushū-shisutemu	学习系统 xuexi xitong
748	least-squares approximation	approximation f des moindres carrés	Approximation durch kleinste Quadrate	аппроксимация методом наименьших квадратов	approssimazione ai minimi quadrati	aproximación f de mínimos cuadrados	最小自乗近似 saisyō-jizyō-kinji	真小二乘逼近 zuixiao ercheng bijin
749	level change value	valeur f de commutation	Sprungwert	пороговая величина	valore di cambiamento di livello	nivel m de conmutación	レベル変更値 reberu-henkōchi	位级切换值 weiji qiehuanzhi
750	lifetime	temps m de vie	Lebensdauer	срок службы	la vita intera	tiempo m de vida	寿命 jumyō	使用期、寿命 shiyongqi, shouming
751	likelihood function	fonction f de vraisemblance	Likelihood-Funktion	функция правдоподобия	funzione di verosimiglianza	función f de verosimilitud	尤度関数 yūdo-kansū	似然函数 siran hanshu
752	limit cycle	cycle m limite	Grenzzyklus, Grenzschwingung	предельный цикл	ciclo limite	ciclo m límite	リミットサイクル rimitto-saikuru	极限环 jixianhuan
753	limitation	limitation f	Begrenzung, Beschränkung	ограничение	limitazione	limitación f	限界 genkai	限制 xianzhi
754	limiter	limiteur m	Begrenzer	ограничитель	limitatore	limitador m	リミタ rimita	限制器 xianzhiqi

58

No.	English	French	German	Russian	Italian	Spanish	Japanese	Chinese
755	limiting control action	action f à commande limitée	begrenzende Regelung (Steuerung)	ограничивающее управляющее воздействие	azione di controllo limitante	acción f de control limitadora	リミット付制御動作	有限控制作用 youxian kongzhi zuoyong
756	limiting feedback	réaction f limitative	begrenzende Rückführung	ограничивающая обратная связь	retroazione limitante	realimentación f limitadora	リミット付フィードバック rimitto-tsuki-fidobakku	有限反馈 youxian fankui
757	limiting feedforward	action f limitative	begrenzende Vorwärtswirkung	ограничивающая прямая связь	azione limitante diretta (in avanti)	acción f anticipante limitadora	リミット付フィードフォード rimitto-tsuki-fidofowādo	有限前馈 youxian qiankui
758	line printer	imprimante f (à lignes)	Zeilendrucker	алфавитно-цифровое печатающее устройство (АЦПУ)	stampante a righe	impresora f de lineas	ラインプリンタ rain-purinta	宽行打印机 kuanhang dayinji
759	linear	linéaire	linear	линейный	lineare	lineal	線形な senkei-na	线性的 xianxingde
760	linear control	régulation f linéaire	lineare Regelung	линейное управление (регулирование)	controllo lineare	control m lineal	線形制御 senkei-seigyo	线性控制 xianxing kongzhi
761	linear dependence	dépendance f linéaire	lineare Abhängigkeit	линейная зависимость	dipendenza lineare	dependencia f lineal	1次従属 ichiji-zyūzoku	线性相关 xianxing youguan
762	linear differential transformer	transformateur-comparateur m	linearer	линейный дифференциальный трансформатор	trasformatore differenziale lineare	transformador m diferencial	線形差動変圧器	线性微分变换 xianxing weifen bianhuan
763	linear filter	filtre m linéaire	lineares Filter	линейный фильтр	filtro lineare	filtro m lineal	線形フィルタ senkei-firuta	线性滤波器 xianxing lüboqi
764	linear independence	indépendance f linéaire	lineare Unabhängigkeit	линейная независимость	indipendenza lineare	independencia f lineal	1次独立 ichiji-dokuritsu	线性独立 xianxing wuguan
765	linear output feedback	rétroaction f linéaire sur la grandeur de sortie	lineare Ausgangsrückführung	линейная обратная связь по выходу	retroazione lineare dall'uscita	realimentación f lineal de salida	線形出力フィードバック senkei-shutsuryoku-fidobakku	线性输出反馈 xianxing shuchu fankui
766	linear programing	programmation f linéaire	lineare Programmierung	линейное программирование	programmazione lineare	programación f lineal	線形計画法 senkei-keikaku-hō	线性规划 xianxing guihua
767	linear system	système m linéaire	lineares System	линейная система	sistema lineare	sistema m lineal	線形系, 線形システム senkei-kei, senkei-shisutemu	线性系统 xianxing xitong

	English	French	German	Russian	Italian	Spanish	Japanese	Chinese
768	linear transformation	transformation f linéaire	Lineartransformation	линейное преобразование	trasformazione lineare	transformación f lineal	線形変換 senkei-henkan	线性变换 xianxing bianhuan
769	linearity	linéarité f	Linearität	линейность	linearità	linealidad f	線形性 senkei-sei	线性 xianxingdu
770	linearization	linéarisation f	Linearisierung	линеаризация	linearizzazione	linealización f	線形化 senkeika	线性化 xianxinghua
771	linearized model	modèle m linéarisé	linearisiertes Modell	линеаризованная модель	modello linearizzato	modelo m linealizado	線形化モデル senkeika-moderu	线性化模型 xianxinghua moxing
772	linkage	mécanisme m de liaison	Verkettung, Verknüpfung	связь, соединение	collegamento	montaje	リンク機構 rinku-kikō	联接，联动装置 lianjie, liandong zhuangzhi
773	linkage multiplier	multiplieur m mécanique	Multipliziergestänge	коэфициент связи	moltiplicatore di collegamento	multiplicador de montaje	リンク乗算器 rinku-jōzanki	联动乘法器 liandong chengfaqi
774	load	charge f	Belastung, Verbraucher, Last (-Widerstand)	нагрузка	carico	carga f	負荷 fuka	负载 fuzai
775	load regulation	statisme m	Lastregelung, Leistungsregelung (Antrieb)	статизм	regolazione del carico	regulación f de carga, estatismo m	負荷変動率 fuka-hendōritsu	负载调节 fuzai tiaojie
776	local area network (LAN)	réseau m local (d'ordinateurs)	lokales Datennetz (LAN)	локальная сеть	rete locale (LAN)	red * de área local	ローカルエリアネットワーク rōkaru-eria-nettowāku	局域网络 juyu wangluo
777	local control (open loop)	commande f locale	lokale Steuerung	местное управление (разомкнутое)	controllo locale (ad anello aperto)	mando m local	局所制御 (開ループ) kyokusho-seigyo (kai-rūpu)	局部控制 (开环) jubu kongzhi (kaihuan)
778	locking	verrouillage m	sperren, verriegeln, arretieren	запирание	blocco, chiusura	enclavamiento m	保持 hoji	锁定 suoding
779	locus	lieu m	Ortskurve, geometrischer Ort	годограф	luogo	lugar m geométrico	軌跡 kiseki	轨迹 guiji
780	logarithmic decrement	décrément m logarithmique	logarithmisches Dekrement	логарифмический декремент	decremento logaritmico	decremento m logarítmico	対数減衰率 taisū-gensuiritsu	对数衰减率 duishu shuaijianlü

	English	French	German	Russian	Italian	Spanish	Japanese	Chinese
781	logarithmic gain	gain m logarithmique	logarithmischer Verstärkungsfaktor	логарифмическое усиление	guadagno logaritmico	ganancia f logarítmica	対数ゲイン taisū-gein	対数増益 duishu zengyi
782	logic analyser	analyseur m de circuits logiques	Logikanalysator	логический анализатор, анализатор логики	analizzatore logico	analizador * lógico	論理解析器 rojikku-anaraiza	逻辑分析器 luoji fenxiqi
783	logic array	réseau m de portes logiques	logische Anordnung	логический ряд	griglia logica	matriz * lógico	ロジックアレイ rojikku-arei	逻辑阵列 luoji zhenlie
784	logic circuit	circuit m logique	logische Schaltung, Logikschaltung	логическая цепь	circuito logico	circuito * lógico	論理回路 ronri-kairo	逻辑线路 luoji xianlu
785	logic design	synthèse f de fonctions logiques, conception f (de la logique)	logischer Entwurf	логическое проектирование, синтез логики	progetto logico	diseño * logico	論理設計 ronri-sekkei	逻辑设计 luoji sheji
786	logic diagram	schéma m logique	logischer Schaltplan, Logikdiagramm	логическая схема	diagramma logico	esquema m lógico	論理線図 ronri-senzu	逻辑图 luojitu
787	logic gate	porte f logique	Verknüpfungsglied	логический вентиль	porta logica	puerta * lógica	論理ゲート ronri-gēto	逻辑门 luojimen
788	logic unit	élément m logique	Logikeinheit	логический блок	unità logica	unidad f lógica	論理ユニット ronri-yunitto	逻辑单元 luoji danyuan
789	logical control	régulation f logique	Verknüpfungssteuerung, logische Steuerung	логическое управление	controllo logico	control m lógico	論理制御 ronri-seigyo	逻辑经制 luoji kongzhi
790	logical operation	opération f logique	logische Verknüpfung	логическая операция	operazione logica	operación f lógica	論理演算 ronri-enzan	逻辑操作 luoji caozuo
791	logical product	produit m logique (normal)	logisches Produkt, logische Multiplikation	конъюнкция, логическое произведение	prodotto logico	producto m lógico	論理積 ronri-seki	逻辑积 luojiji
792	logical sum	somme f logique (normale)	logische Summe, logische Summenbildung	дизъюнкция, логическая сумма	somma logica	suma f lógica	論理和 ronri-wa	逻辑和 luojihe
793	loop	boucle f	Regelkreis, Schleife, Kreis	контур	anello	anillo m	ループ rūpu	回路 huilu

No.	English	français	Deutsch	русский	italiano	español	日本語	中文
794	loop element	organe *m* de boucle	Regelkreiselement	элемент контура	elemento dell'anello	elemento *m* del anillo	ループ要素 rūpu-yōso	回路环节，回路元件 huilu huanjie, huilu yuanjian
795	loop gain	gain *m* en boucle ouverte	Kreisverstärkung, Verstärkung des offenen Kreises	коэффициент усиления разомкнутого контура	guadagno d'anello	ganancia *f* en anillo abierto	ループゲイン rūpu-gein	回路增益 huilu zengyi
796	loop phase angle	déphasage *m* en boucle ouverte	Phasenwinkel des offenen Regelkreises	фазовый сдвиг разомкнутого контура	sfasamento d'anello	desfase *m* en anillo abierto	ループ位相角 rūpu-isōkaku	回路相位角 huilu xiangweijiao
797	loop transfer	transfert *m* de boucle	offener Regelkreis	контурная передача, передача контура	(funzione di) trasferimento d'anello	transferencia *f* de anillo abierto	ループ伝達 rūpu-dentatsu	回路传递 huilu chuandi
798	loop transfer locus	lieu *m* de transfert	Ortskurve des offenen Regelkreises	диаграмма Найквиста	diagramma della funzione di trasferimento d'anello	diagrama *m* polar de la función de transferencia en anillo abierto	一巡伝達軌跡 ichijun-dentatsu-kiseki, ループ伝達軌跡 rūpu-dentatsu-kiseki	回路轨迹 huilu guiji
799	loss	affaiblissement *m*	Verlust	ослабление, затухание	perdita	atenuación *f*	損失 sonshitsu	损耗 sunhao
800	low level language	langage *m* élaboré, de bas niveau	Grundsprache, niedere Maschinensprache	язык низкого уровня	linguaggio di basso livello	lenguaje *m* de bajo nivel	低レベル言語 teireberu-gengo	低级语言 diji yuyan
801	low-pass filter	filtre *m* passe-bas	Tiefpaßfilter	фильтр (пропускания) нижних частот	filtro passa-basso	filtro *m* pasa-bajo	低域フィルタ tei-iki-firuta	低通滤波器 ditong lüboqi
802	lumped constant model	modèle *m* à paramètres localisés	konzentriertparametrisches Modell	модель с постоянными сосредоточенными параметрами	modello costante aggregato	modelo invariante de parámetros localizados (concentrados)	集中定数モデル shūchū-jyōsū-moderu	集总常数模型 jizong changsu moxing
803	lumped model	modèle *m* localisé	konzentriertes Modell	модель с сосредоточенными параметрами	modello aggregato	modelo *m* de parámetros localizados	集中化モデル shūchūka-moderu	集总模型 jizong moxing
804	lumped-parameter system	système *m* à paramètres localisés	konzentriert-parametrisches System	система с сосредоточенными параметрами	modello a parametri concentrati	sistema *m* de parámetros localizados	集中パラメータ系 shūchū-paramēta-kei	集总参数系统 jizong canshu xitong
805	Lyapunov function	fonction *f* de Lyapounov	Ljapunov-Funktion	функция Ляпунова	funzione di Lyapunov	función *f* de Lyapunov	リアプノフ関数 Lyapunov-kansū	李雅普诺夫函数 liyapunuofu hanshu
806	Lyapunov method	méthode *f* de Lyapounov	Ljapunov-Methode	метод Ляпунова	metodo di Lyapunov	método *m* de Lyapunov	リアプノフ法 Lyapunov-hō	李雅普诺夫方法 liyapunuofu fangfa

English	French	German	Russian	Italian	Spanish	Japanese	Chinese	No.
Lyapunov stability	stabilité f au sens de Lyapounov	Ljapunov-Stabilität	устойчивость по Ляпунову	stabilità alla Lyapunov	estabilidad f según Lyapunov	リアプノフ安定性 Lyapunov-anteisei	李雅普诺夫稳定性 liyapunuofu wendingxing	807
Lyapunov vector function	fonction f vectorielle de Lyapounov	Ljapunovsche Vektorfunktion	векторная функция Ляпунова	funzione vettoriale di Lyapunov	función f vectorial de Lyapunov	リアプノフベクトル関数 Lyapunov-bekutoru-kansū	李雅普诺夫向量函数 liyapunuofu xiangliang hanshu	808
machine code	langage m, code m machine	Maschinencode	машинный код	codice macchina	código * de máquina	マシーンコード mashin-kōdo	机器代码 jiqi daima	809
machine language	langage m machine	Maschinensprache	машинный язык	linguaggio macchina	lenguaje m máquina	機械語 kikaigo	机器语言 jiqi yuyan	810
magnetic amplifier	amplificateur m magnétique	Magnetverstärker	магнитный усилитель	amplificatore magnetico	amplificador m magnético	磁気増幅器 jiki-zōfukuki	磁放大器 cifangdaqi	811
magnetic brake	frein m magnétique	Magnetbremse	(электро)магнитный тормоз	freno magnetico	freno magnético	電磁ブレーキ denji-burēki	磁力制动 cili zhidong	812
magnetic clutch	embrayage m magnétique	Magnetkupplung	(электро)магнитная муфта	innesto magnetico	embrague magnético	電磁クラッチ 磁気クラッチ denji-kuratti jiki-kuratti	电磁离合器 dianci liheqi	813
magnetic fluid clutch	embrayage m à fluide magnétique	Magnet-Flüssigkeitskupplung	магнитная жидкостная муфта	innesto a fluido magnetico	embrague por fluido magnético	磁気流体クラッチ jiki-ryūtai-kuratti	磁性流体离合器 cixing liuti liheqi	814
magnetic modulator	modulateur m magnétique	Magnetmodulator	магнитный модулятор	modulatore magnetico	modulador m magnético	磁気変調器 jiki-henchōki	磁调制器 citiaozhiqi	815
magnetic powder clutch	embrayage m à poudre magnétique	Magnetpulverkupplung	магнитная порошковая муфта	innesto a polvere magnetica	embrague por poluo magnético	磁気粉末クラッチ jiki-funmatsu-kuratti	磁粉离合器 cifen liheqi	816
magnification, Q	facteur m de résonance	Resonanzüberhöhung	относительный резонансный пик	ingrandimento, fattore di merito Q	coeficiente m de resonancia	共振の強さ kyōshin-no-tsuyosa	放大, 放大率 fangda, fangdalü	817
magnitude contour	courbe f d'amplitude	Betragsverlauf	линия постоянного козффициента усиления	curva di livello del modulo	curva f de amplitud	振幅軌跡 shinpuku-kiseki	等幅线 dengfuxian	818
main memory	mémoire f principale	Hauptspeicher	оперативное заломинающее устройство (ОЗУ), оперативная память	memoria principale	memoria f principal	主記憶 shu-kioku	主存储器 zhucunchuqi	819

63

	English	French	German	Russian	Italian	Spanish	Japanese	Chinese
820	major loop	boucle f principale	Hauptkreis	основной контур	anello primario	anello m principal	主要ループ shuyō-rūpu	主回路 zhuhuilu
821	majority logic	logique f majoritaire	Mehrheitslogik	мажорантная логика, мажоритарная логика, логика большинства	logica a maggioranza	lógica * de mayoría	多数決論理 tasuketsu-ronri	多数逻辑 duoshu luoji
822	man/machine interaction	interaction f homme/machine	Mensch-Maschine Interaktion	общение человека с машиной	interazione uomo/macchina	interacción f hombre-máquina	人間／機械一相互干渉 ningen-kikai-sōgokanshō	人-机交互 ren-ji jiaohu
823	man/machine interface	interface f homme/machine	Mensch-Maschine Schnittstelle	сопряжение человека с машиной, человеко-машинный интерфейс	interfaccia uomo/macchina	interfase f hombre-máquina	人間／機械一インターフェイス ningen-kikai-intāfeisu	人-机界面 ren-ji jiemian
824	man-machine system	système m homme/machine	Mensch-Maschine System	человеко-машинная система	sistema uomo/macchina	sistema m hombre-máquina	人間／機械系 ningen-kikai-kei	人-机系统 ren-ji xitong
825	manipulated variable	grandeur f de commande	Stellgröße	преобразованная переменная	variabile manipolata	magnitud f manipulada	制御量 seigyo-ryō	被操纵变量, 被控变量 beicaozong bianliang, beikong bianliang
826	manipulator	manipulateur m	Handhabungsgerät, Manipulator	манипулятор	manipolatore	manipulador m	マニピュレータ manipyurēta	机械手 jixieshou
827	manual	manuel m	Hand, Handbetätigung, Handbuch	ручной	manuale	manual	手動の shudō-no	手动的, 人工操作的 shoudongde, rengong caozuode
828	manual closed-loop control system	système m de commande manuelle à asservissement	Handregelsystem	ручное управление в замкнутой системе	sistema di controllo manuale ad anello chiuso	sistema m manual de control en anillo cerrado	手動閉路制御系 shudō-heikairo-seigyokei	手动闭环控制系统 shoudong bihuan kongzhi xitong
829	manual control (open loop)	commande f manuelle	Handsteuerung (offener Kreis)	ручное управление (разомкнутое)	controllo manuale (ad anello aperto)	control m manual (anillo abierto)	手動制御 shudō-seigyo	手动控制 (开环) shoudong kongzhi (kaihuan)
830	manual monitored control system	système m de régulation supervisé manuellement	handbetätigtes Regelungs-(Steuerungs-)System	ручное управление в замкнутой системе	sistema di controllo a presidio manuale	sistema m manual de control en anillo cerrado, sistema m manual de control monitorizado	手動監視制御系	手动监控系统 shoudong jiankong xitong
831	manual operation	fonctionnement m manuel	Handbetätigung, Betätigung von Hand	ручное действие	operazione manuale	operación f manual, funcionamiento m manual	手動操作 shudō-sōsa	手动操作 shoudong caozuo
832	many-degrees-of-freedom system	système m à plusieurs degrés de liberté	System mit vielen Freiheitsgraden	система со многими степенями свободы	sistema a molti gradi di libertà	sistema m con varios grados de libertad	多自由度系 ta-jiyūdo-kei	多自由度系统 duoziyoudu xitong

No.	English	French	German	Russian	Italian	Spanish	Japanese	Chinese
833	marginal stability	stabilité f marginale	marginale Stabilität	предельная устойчивость, маргинальная устойчивость	stabilità marginale	estabilidad f marginal	限界安定性 genkai-anteisei	临界稳定性 linjie wendingxing
834	mark/space ratio	"rapport m des états 1 aux états 0" dans un message	Impuls-Pausenverhältnis	отношение уровень/объём	rapporto segno/spazio	relación * marca/blanco	マーク/スペース・比 māku/supēsu-hi	信一空比 xin-kong bi
835	Markov chain	suite f markovienne	Markovkette	цепь Маркова, марковская цепь	catena di Markov	cadena de Markov	マルコフ連鎖 Markov-rensa	马尔可夫链 maerkefu lian
836	Markov process	processus m markovien	Markovprozeß	марковский процесс, процесс Маркова	processo di Markov	proceso Markoviano	マルコフ過程 Markov-katei	马尔可夫过程 maerkefu guocheng
837	Markovian system	système m markovien	Markovsches System	марковская система	sistema markoviano	sistema m Markoviano	マルコフシステム Markov-shisutemu	马尔可夫系统 maerkefu xitong
838	master station	poste m maître	Hauptstation	ведущая станция	stazione principale	estación f maestra, estación f principal	主 局	主站 zhuzhan
839	matched filter	filtre m adapté	angepaßter Filter	согласованный фильтр	filtro adattato	filtro * adaptado		匹配滤波器 pipei luboqi
840	material-balance control	régulation f d'équilibre de masses	Materialbalance-Regler	управление материальным балансом	controllo a bilancio di materiali	control m del balance de materiales	物質平衡制御 busshitsu-heikō-seigyo	材料平衡控制 cailiao pingheng kongzhi
841	mathematical model	modèle m mathématique	mathematisches Modell	математическая модель	modello matematico	modelo m matemático	数学モデル sūgaku-moderu	数学模型 shuxue moxing
842	matrix	matrice f	Matrix	матрица	matrice	matriz f	行列 gyōretsu	矩阵 juzhen
843	matrix algebra	algèbre f matricielle	Matrizenalgebra	матричная алгебра	algebra delle matrici	álgebra * matricial	行列演算 gyōretsu-enzan	矩阵代数 juzhen daishu
844	matrix determinant	déterminant m (d'une matrice)	Determinante einer Matrix	матричный детерминант, определитель	determinante di una matrice	determinante * de la matriz	行列式 gyōretsu-shiki	矩阵行列式 juzhen hanglieshi
845	matrix element	élément m de matrice	Element einer Matrix, Matrizenelement	элемент матрицы	elemento di una matrice	elemento * de la matriz	行列要素 gyōretsu-yōso	矩阵元素 juzhen yuansu

No.	English	French	German	Russian	Italian	Spanish	Japanese	Chinese
846	matrix printer	imprimante f matricielle	Matrixdrucker	матричный принтер	stampante a matrice	matriz f impresora (matriz de puntos)	ドットプリンタ dotto-purinta	矩阵式打印机 juzhenshi dayinji
847	maximization	maximisation f	Maximierung	максимизация	massimizzazione	maximización f	最大化 saidaika	极大化 jidahua
848	maximize (v)	maximiser (v)	maximieren	максимизировать	massimizzare	maximizar	最大化する saidaika-suru	极大化 (动词) jidahua (dongci)
849	maximum principle (of Pontryagin)	théorème m du maximum, de Pontryagine	(Pontrjaginsches) Maximumprinzip	принцип максимума (Понтрягина)	principio del massimo (di Pontryagin)	principio * del máximo (de Pontryagin)	Pontryaginの最大原理 Pontryagin-no-saidai-genri	(庞特里亚金)最大值原理 (pangteliyajin) jidazhi yuanli
850	mean	moyenne f	Mittelwert	среднее (значение)	media	media	平均 heikin	平均值 pingjunzhi
851	mean-square error	écart m, erreur f quadratique moyen(ne)	mittlerer quadratischer Fehler	среднеквадратичная ошибка	errore quadratico medio	error * cuadrático medio	自乗平均誤差 jijō-heikingosa	均方误差 junfang wucha
852	mean time between failures (MTBF)	temps m moyen entre défaillance (T.M.E.D.)	mittlerer Ausfallabstand (MTBF)	наработка между отказами, наработка на отказ	tempo medio fra due guasti (MTBF)	tiempo m medio entre fallos	MTBF, 平均故障間隔 heikin-koshō-kankaku	平均无障碍时间 pingjun guzhang jian'ge shijian
853	mean time to repair (MTTR)	temps m moyen de réparation (T.M.R.)	mittlere Reparaturdauer (MTTR)	среднее время восстановления, среднее время ремонта	tempo medio di riparazione (MTTR)	tiempo m medio de reparación	MTTR, 平均修理時間 heikin-shūri-jikan	平均修复时间 pingjun xiufu shijian
854	mean time to failure (MTTF)	temps m moyen avant défaillance (T.M.A.D.)	mittlere Zeit bis zum ersten Ausfall (MTTF)	наработка до отказа	tempo medio di guasto (MTTF)	tiempo m medio sin fallo	MTTF, 平均寿命 heikin-jumyō	平均无故障时间 pingjun wuguzhang shijian
855	measured feedback	réaction f mesurée	gemessene Rückführgröße	измеренная обратная связь	retroazione misurata	realimentación f medida	測定 (株出) フィードバック sokutei-fidobakku	测量反馈 celiang fankui
856	measured value	valeur f mesurée	Meßwert	измеренное значение	valore misurato	valor m medido	測定値 sokutei-chi	测量值 celiangzhi
857	measuring element	organe m de mesure	Meßeinrichtung	измерительный элемент	elemento di misura	elemento m de medición, medidor m	測定 (株出) 要素 sokutei-(kenshutu)-yōso	测量元件 celiang yuanjian
858	measuring point	lieu m de mesure, emplacement m de mesure	Meßpunkt	точка измерения	punto di misura	punto de medición	測定点 sokutei-ten	测量点 celiangdian

No.	English	French	German	Russian	Italian	Spanish	Japanese	Chinese
859	measuring range	plage f de mesure	Meßbereich	диапазон измерения	gamma di misura	margen m de medición	測定範囲 sokutei-han-i	測量范围 celiang fanwei
860	measuring span	étendue f de mesure	Meßspanne	диапазон измерения, интервал измерения	ampiezza di misura	margen m de medida	スパン supan	量程 liangcheng
861	measuring transducer	transducteur m de mesure	Meßwertumformer, Meßumformer	датчик измерения, измерительный преобразователь	trasduttore di misura	transductor m de medida	検出用トランスデューサ kenshutsuyō-toransudyūsa	测量传感器 celiang chuan'ganqi
862	measuring transmitter	transducteur m de mesure	Meßwertüberträger	датчик, измерительный преобразователь	trasmettitore di misura	transmisor m de medida	測定用発信器 (トランスデューサ) sokuteiyo-hasshinki (toransu-dyūsa)	测量变送器 celiang biansongqi
863	measuring unit	capteur m de mesure, transmetteur m de mesure	Meßeinheit, Meßeinrichtung	измерительное устройство	unità f di misura	unidad f de medida	検出部 kenshutsu-bu	测量单元 celiang danyuan
864	mechanization	mécanisation f	Mechanisierung	механизация	meccanizzazione	mecanización f	機械化 kikai-ka	机械化 jixie hua
865	memory	mémoire f	Speicher, Gedächtnis	память	memoria	memoria f	記憶 メモリ kioku memori	存储器 cunchuqi
866	memory bank	batterie f de mémoires	Speicherbank	банк памяти	banco di memoria	banco * de memoria	メモリーバンク memori-banku	存储器库 cunchuqi ku
867	messmotor	moteur m de mesure	Meßmotor	мессмотор	'messmotor'			积分电动机 jifen diandongji
868	metadyne	métadyne f	Metadyne, Konstantenstromverstärkermaschine	метадин	metadinamo	metadina	メタダイン metadain	交磁电机放大器 jiaoci dianji fangdaqi
869	metadyne generator	dynamo f métadyne	Metadynegenerator, Querfeldgenerator	генератор метадин	metadinamo (generatrice)	generador de metadina	メタダイン発電器 metadain-hatsudenki	交磁电机放大器发电机 jiaoci dianji fangdaqi fadianji
870	method of control	mode m de commande	Regelungs-(od. Steuerungs-)Methode	принцип управления	metodo di controllo	tipo m de control	制御法 seigyo-hō	控制方法 kongzhi fangfa
871	method of duality	méthode f par principe de dualité	Dualitätsprinzip	метод дуальности	metodo di dualità	método m de la dualidad	双対法 sotsui-hō	对偶方法 dui'ou fangfa

	English	French	German	Russian	Italian	Spanish	Japanese	Chinese
872	method of least squares	méthode f des moindres carrés	Methode der kleinsten Quadrate	метод наименьших квадратов	metodo dei minimi quadrati	método m de los mínimos cuadrados	最小自乗法 saishōjijo-hō	最小二乘法 zuixiao erchengfa
873	method of order reducing	méthode f de réduction d'ordre	Ordnungsreduzierungsverfahren	метод понижения порядка	metodo di riduzione dell'ordine	método m de reducción de orden	低次元化法 teijigenka-hō	降阶法 jiangjiefa
874	method of perturbations	méthode f de perturbations	Störungsmethode, Störungsrechnung	метод возмущений	metodo delle perturbazioni	método m de perturbaciones	摂動法 setsudō-hō	摄动法 shedongfa
875	method of regression	méthode f de régression	Regressionsmethode	регрессионный метод	metodo di regressione	método m de regresión	回帰モデル法 kaikimoderu-hō	回归法 huiguifa
876	method of variation of the constants (or parameters)	méthode f de variation des constantes	Methode der Variation der Konstanten	метод вариации постоянных (или параметров)	metodo di variazione delle costanti (o dei parametri)	método m de variación de las constantes	定数変化法 jōsuhenka-hō	系数(或参数)变化法 xishu (huocanshu) bianhuafa
877	method of weighted residuals	méthode f des résidus pondérés	Methode der gerichteten Residualen	метод взвешенных остатков	metodo dei residui pesati	método m de residuos ponderados	重み付き残差法 omomitsuki-zansa-hō	加权剩余法 jiaquan shengyufa
878	microcomputer	micro-ordinateur m, micro-calculateur m	Mikrorechner, Mikrocomputer	микрокомпьютер	microcalcolatore	sistema m con microprocesador	マイクロコンピュータ maikuro-konpyūta	微型计算机 weixing jisuanji
879	microcomputer system	système m à micro-calculateur	Mikrorechnersystem	микрокомпьютерная система	sistema a microcalcolatore	microprocesador m	マイクロコンピュータシステム maikuro-konpyūta-shisutemu	微型计算机系统 weixing jisuanji xitong
880	microprocessor	micro-processeur m	Mikroprozessor (MP)	микропроцессор	microprocessore	micoprocesador	マイクロプロセッサ maikuro-purosessa	微处理机 weichuliji
881	microprograming	micro-programmation f	Mikroprogrammierung, Mikroprogrammsteuerung	микропрограммирование	microprogrammazione	microprogramación *	マイクロプログラミング maikuro-puroguramingu	微程序设计 weichengxu sheji
882	minicomputer	mini-ordinateur m, mini-calculateur m	Minicomputer, Kleinrechner	миникомпьютер, мини-ЭВМ	minicalcolatore	minicomputador *	ミニコンピュータ mini-konpyūta	小型计算机 xiaoxing jisuanji
883	minimization	minimisation f	Minimierung	минимизация	minimizzazione	minimización f	最小化 saishōka	极小化 jixiaohua
884	minimize (v)	minimiser (v)	minimieren	минимизировать	minimizzare	minimizar	最小化する saishōka-suru	使极小化 (动词) shijixiaohua (dongci)

No.	English	French	German	Russian	Italian	Spanish	Japanese	Chinese
885	minimum phase-shift system	système m à déphasage minimal	Phasenminimumsystem, System	минимально-фазовая система	sistema a sfasamento minimo	sistema m de desfase mínimo	最小位相推移系 saishō-isō-suii-kei	最小相移系统 zuixiao xiangyi xitong
886	minimum-time control	régulation f brachystochrone	zeitminimale Regelung	управление по (максимальному) быстродействию, управление с наименьшим временем	controllo in tempo minimo	control m de tiempo mínimo	最小時間制御 saishō-jikan-seigyo	最小时间控制 zuixiao shijian kongzhi
887	minimum trajectory	trajectoire f minimale	minimale Trajektorie	минимальная траектория	traiettoria minima	trayectoria f minima	最小軌道 saishō-kidō	最小轨迹 zuixiao guiji
888	minor loop	boucle f auxiliaire	Nebenkreis, Unterkreis, Nebenschleife	вспомогательный контур	anello secondario	anillo m secundario	局所ループ kyokusho-rūpu	副回路 fuhuilu
889	misalignment	écart m	schlechte Einstellung, Fehlausrichtung, Fluchtungsfehler	разрегулированность, угловое рассогласование	disallineamento	desalineación f, desviación f angular	不整合	未对准 weiduizhun
890	modal control	régulation f modale	modale Regelung	модальное управление	controllo modale	control m modal	モード制御 mōdo-seigyo	模态控制 motai kongzhi
891	modal system	système m modal	modales System	модальная система	sistema modale	sistema modal	モードシステム mōdo-shisutemu	模态系统 motai xitong
892	modal transformation	transformation f modale	Transformation auf modale Darstellung	модальное преобразование	trasformazione modale	transformación f modal	モード変換 mōdo-henkan	模态变换 motai bianhuan
893	modeling	modélisation f	Modellierung	моделирование	modellistica	modelado m	モデル化 moderu-ka	建模 jianmo
894	modem	modem m, modulateur-démodulateur m	Modem (Modulator/Demodulator)	модем	modem	módem *	モデム modemu	调制解调器 tiaozhi jietiaoqi
895	modifying feedback	réaction f réglable	modifizierende Rückführung	модифицирующая обратная связь	retroazione modificatrice	realimentación f modificadora	修正フィードバック shūsei-fidobakku	修正反馈 xiuzheng fankui
896	modifying feedforward	action f directe réglable	modifizierende Steuerung (od. Vorwärtswirkung)	модифицирующая прямая связь	azione correttiva ad anello aperto	alimentación f modificadora	修正フィードフォワード shūsei-fidofowādo	修改前馈 xiugai qiankui
897	modulating action	action f par modulation	Modulationsverfahren	модуляционное действие	azione modulante	acción f moduladora	変調動作 henchō-dōsa	调制作用 tiaozhi zuoyong

No.	English	français	Deutsch	русский	italiano	español	日本語	中文
898	modulation	modulation f	Modulation, Aussteuerung	модуляция	modulazione	modulación f	変調 henchō	调制 tiaozhi
899	modulator	modulateur m	Modulator	модулятор	modulatore	modulador m	変調器 henchōki	调制器 tiaozhiqi
900	moment of inertia	moment m d'inertie	Trägheitsmoment	момент инерции	momento d'inerzia	momento m de inercia	慣性モーメント kansei-mōmento	转动惯量、惯性矩 zhuandong guanliang, guanxingju
901	monitored control system	système m de régulation supervisé	Regelsystem (G.B.)	замкнутая система управления	sistema di controllo vigilato	sistema m de control en anillo cerrado, sistema m monitorizado de control	監視付制御系 kanshi-tsuki-seigyokei	监控系统 jiankong xitong
902	monitoring	surveillance f	Überwachung, überwachen	наблюдение, контролирование, мониторинг	sorveglianza (o controllo)	monitorización f	監視 kanshi	监控、监测 jiankong, jiance
903	monitoring element	organe m de réaction principale	Hauptwirkungsglied	контрольный элемент	elemento di sorveglianza	elemento m de monitorización, elemento m de realimentación principal	監視要素 kanshi-yōso	监控元件 jiankong yuanjian
904	monitoring feedback	réaction f principale	Hauptrückführung Kontrollrückführung	контролирующая обратная связь	retroazione di sorveglianza (supervisione e controllo)	realimentación f de monitorización, realimentación f principal	監視フィードバック kanshi-fīdobakku	监控反馈 jiankong fankui
905	monitoring loop	boucle f de réaction principale	Hauptrückführkreis	контур управления	anello di sorveglianza (supervisione e controllo)	anillo m de monitorización, anillo m de realimentación principal	監視ループ kanshi-rupu	监控回路 jiankong huilu
906	monostable multivibrator	multivibrateur m monostable	monostabiler Multivibrator, Kippglied	моностабильный мультивибратор, одновибратор	multivibratore monostabile	multivibrador m monoestable	単安定マルチバイブレータ tan-antei-maruchibaibureta	单稳多谐振荡器 danwen duoxie zhendangqi
907	monostable trigger element	basculeur m monostable	monostabiles Kippglied	моностабильный триггер	'trigger' (o generatore comandato di segnale d'avvio) monostabile	(elemento) disparador m monoestable	単安定トリガ要素 tan-antei-toriga-yōso	单稳触发元件 danwen chufa yuanjian
908	Monte Carlo method	méthode f de Monte Carlo	Monte Carlo-Methode	метод Монте-Карло	metodo Monte Carlo	método m de Montecarlo	モンテカルロ法 Monte Carlo-hō	蒙特卡罗法 mengtekaluofa
909	motor	moteur m	Motor	двигатель, мотор	motore	motor m	モーター dendōki	马达、电动机 mada, diandongji
910	motor element	organe m motorisé	Motorteilsystem	приводной элемент, моторный элемент	elemento motore	elemento m motor	操作要素 sōsa-yōso	马达元件 mada yuanjian

No.	English	French	German	Russian	Italian	Spanish	Japanese	Chinese
911	multiaccess system	système m à grandeurs d'entrée multiples	Mehrfachzugriffssystem	система с множественным доступом	sistema multiaccesso	sistema m multiacceso	多重アクセスシステム taijū-akusesu-shisutemu	多路存取系统 duolu cunqu xitong
912	multiprograming	multiprogrammation m	Mehrprogrammbetrieb, Multiprogrammierung	мультипрограмми-рование	multiprogrammazione	multiprogramación *	多重プログラミング taijū-puroguramingu	多道程序设计 duodao chengxu sheji
913	multiaction controller	régulateur m à actions multiples	Mehrfunktionsregler	регулятор множественного воздействия	controllore ad azione multipla	controlador m de acción múltiple	多動作制御装置	多作用控制器 duozuoyong kongzhiqi
914	multichannel controller	régulateur m à canaux multiples	Mehrkanalregler	многоканальный регулятор	controllore a più canali	controlador m multicanal	多チャンネル制御装置 tachanneru-seigyo-sōchi	多通道控制器 duotongdao kongzhiqi
915	multidimensional system	système m multidimensionnel	mehrdimensionales System	многомерная система	sistema multidimensionale	sistemas * multidimensional	多次元システム tajigen-shisutemu	多维系统 duowei xitong
916	multi-input/multi-output (MIMO) system	système m à grandeurs d'entrée et de sortie multiples	Mehrgrößensysteme (MIMO)	система с многими входами-выходами	sistema a più ingressi e più uscite (MIMO)	sistema m de entradas y salidas múltiples	多入力/多出力系 tanyūryoku/tashutsuryoku-kei	多输入多输出系统 duoshuru duoshuchu xitong
917	multilayer structure	structure f à couches multiples, feuilletée, stratifiée	Mehrschichtstruktur, Mehrlagenstruktur	многослойная структура	struttura a più strati	estructura f multicapa	多層構造 tasō-kōzō	多层结构 duoceng jiegou
918	multilevel action	action f à niveaux multiples	Mehrschichtsystem, Mehrpunktverhalten (G.B.)	многопозиционное воздействие	azione a più livelli	acción f multinivel	多層レベル動作 tajūreberu-dōsa	多级作用 duoji zuoyong
919	multilevel control	régulation f à niveaux multiples	Mehrebenenregelung	многоуровневое управление	controllo a più livelli	control m multinivel	階層制御 kaisō-seigyo	多级控制 duoji kongzhi
920	multilevel controller	régulateur m à action à échelons	Mehrebenenregler, Mehrpunktregler (G.B.)	многопозиционный регулятор	controllore a più livelli	controlador m multinivel	階層制御装置 kaisō-seigyo-sōchi	多级控制器 duoji kongzhiqi
921	multilevel structure	structure f à niveaux multiples	Mehrebenenanordnung	многоуровневая структура	struttura a più livelli	estructura f multinivel	多層構造 tasō-kōzo	多级结构 duoji jiegou
922	multilevel system	système m à niveaux multiples	Mehrebenensystem	многоуровневая система	sistema a più livelli	sistema m multinivel	多層システム taijū-shisutemu	多级系统 duoji xitong
923	multiloop control	régulation f à boucles multiples	mehrschleifige Regelung	многоконтурное управление	controllo a più anelli (di retroazione)	control m con anillos múltiples	多重ループ制御 taijū-rūpu-seigyo	多回路系统 duohuilu xitong

№	English	français	Deutsch	Русский	italiano	español	日本語	中文
924	multiplexer	multiplexeur m	Multiplexer, Mehrfachkoppler	мультиплексор	multiplatore	multiplexor m	マルチプレックサ 多重化器 maruchipurekkusa tajūkaki	多路转接器 duolu zhuanjieqi
925	multiposition controller	régulateur m à positions multiples	Mehrpunktregler	многопозиционный регулятор	controllore a pió posizioni	controlador m multiposición	多値制御装置 tachi-seigyo-sōchi	多位置控制 duoweizhi kongzhi
926	multiprocessing system	traitement m multidimensionnel	Mehrprozessorsystem, Mehrrechnersystem	многопроцессорная система, система мультиобработки	sistema ad elaborazione multipla	sistemas * de multiprocesamiento	多重プロセッシングシステム taju-purosesshingu-shisutemu	多重处理系统 duochong chuli xitong
927	multispeed controller	régulateur m à vitesse d'action multiple	Regler mit verschiedenen Geschwindigkeiten	многоскоростной регулятор	controllore a pió velocità	controlador m de velocidades múltiples	多速度調節器 tasokudo-chōsetsuki	多速控制器 duosu kongzhiqi
928	multispeed floating action	action f flottante à plusieurs vitesses	Regler mit verschiedenen Integrationsge-schwindigkeiten	многоскоростное астатическое воздействие	azione flottante a pió velocità	acción flotante multivelocidad	多速浮动動作 tasokudo-fudō-dōsa	多速浮动作用 duosu fudong zuoyong
929	multistep action	action f à échelons multiples	Mehrpunktverhalten	многопозиционное воздействие	azione a passi multipli	acción f escalonada múltiple	多値動作 tachi-dōsa	多级作用 duoji zuoyong
930	multistep controller	régulateur m à échelons multiples	Mehrpunktregler, Schrittregler	многопозиционный регулятор	controllore a passi multipli	controlador m escalonado múltiple	多値調節器 tachi-chōsetsuki	多级控制器 duoji kongzhiqi
931	multivariable control	régulation f à grandeurs d'entrée et de sortie multiples	Mehrgrößenregelung	многосвязное управление, многомерное управление	controllo multivariabile	control m multivariable	多変数制御 tahensū-seigyo	多变量控制 duobianliang kongzhi
932	multivariable control system	système m de régulation à grandeurs d'entrée et de sortie multiples	Mehrgrößenregelsystem	многомерная (=многосвязная) система управления	sistema di controllo multivariabile	sistema m de control multivariable	多変数制御系 (システム) tahensū-seigyo-kei (-shisutemu)	多变量控制系统 duobianliang kongzhi xitong
933	multivariable feedback control	asservissement m à grandeurs d'entrée et de sortie multiples	Mehrgrößenregelung	многосвязное управление с обратной связью	controllo multivariabile a retroazione	control m multivariable con realimentación negativa	多変数フィードバック制御 tahensū-fidobakku-seigyo	多变量反馈控制 duobianliang fankui kongzhi
934	multivariable system	système m à grandeurs d'entrée et de sortie multiples	Mehrgrößensystem	многомерная (=многосвязная) система	sistema multivariabile	sistema m multivariable	多変数系 (システム) tahensū-kei (-shisutemu)	多变量系统 duobianliang xitong
935	multivibrator	multivibrateur m	Multivibrator	мультивибратор	multivibratore	multivibrador m	マルチバイブレータ maruchibaiburēta	多谐振荡器 duoxie zhendangqi
936	NAND element	circuit m ON, circuit m NON-ET	NAND-Glied	элемент И-НЕ	elemento NAND	elemento m NO-Y	ナンド要素 nando-yōso	"与非"元件 "yufei" yuanjian

	English	French	German	Russian	Italian	Spanish	Japanese	Chinese
937	NAND operation	opération f ON, opération f NON-ET	NAND-Verknüpfung (od. Operation)	операция И-НЕ	operazione NAND	operación f NO-Y	ナンド演算 nando-enzan	"与非"运算, "与非"操作 "yufei" yunsuan, "yufei" caozuo
938	natural frequency	fréquence f propre	Eigenfrequenz	собственная частота	frequenza (o pulsazione) naturale	frecuencia f propia	固有周波数 koyū-shūhasū	自然频率 ziran pinlü
939	navigation system	système m de navigation	Navigationssystem	система навигации	sistema di navigazione	sistema * de navegación	航法装置 kōhō-sōchi	导航系统 daohang xitong
940	negation	négation f	Negation, Verneinung	отрицание	negazione	negación f	否定 hitei	非, "非"运算 fei, "fei" yunsuan
941	negative feedback	réaction f négative	negative Rückführung	отрицательная обратная связь	retroazione negativa	realimentación f negativa	負のフィードバック fu-no-fidobakku	负反馈 fufankui
942	network	réseau m à retard	Netzwerk, Netz	сеть	rete	red f	回路網 kairo-mō	网络 wangluo
943	network analyser	analyseur m de réseau	Netzmodell, Netzwerkanalysator	анализатор сетей	analizzatore di rete	analizador m de redes	回路網解析機 kairomō-kaisekiki	网络分析器 wangluo fenxiqi
944	network topology	topologie f du réseau	Netztopologie	топология сети	topologia della rete	topología * de la red	ネットワークトポロジ nettowāku-toporoji	网络拓扑 wangluo tuopuxue
945	neutral zone	zone f d'insensibilité	neutrale Zone, Nullzone, Totzone	зона нечувствительности, нейтральная зона	zona morta (o neutra)	zona f de insensibilidad, zona f muerta	中立帯 chūritsutai	中性区 zhongxingqu
946	neutral zone control	régulation f à zone d'insensibilité	Regelung mit Unempfindlichkeit-sbereich	управление с зоной нечувствительности	controllo a zona morta (o neutra)	control de zona muerta		中性区控制 zhongxingqu kongzhi
947	Nichols chart	diagramme m de Nichols	Nichols-Diagrammblatt	диаграмма Николса	carta di Nichols	diagrama m de Nichols	ニコルス線図 Nichols-senzu	尼科尔斯图 nikeersitu
948	Nichols diagram	diagramme m de Black-Nichols	Nichols-Diagramm	диаграмма Николса	diagramma di Nichols	diagrama m de Nichols	ニコルス線図 Nichols-senzu	尼科尔斯图 nikeersitu
949	node	noeud m	Knoten, Knotenpunkt	узел	nodo	nodo m	結節点 kessetsuten	节点, 结点 jiedian, jiedian

No.	English	French	German	Russian	Italian	Spanish	Japanese	Chinese
950	noise	bruit m	Rauschen, Lärm, Geräusch	шум	rumore	operación f NO-Y	ノイズ 雑音 noizu zatsuon	噪声 zaosheng
951	noise level	niveau m de bruit	Rauschpegel, Störpegel, Geräuschstärke	уровень шумов, уровень помех	livello di rumore	frecuencia f propia	ノイズレベル noizu-reberu	噪声水平 zaosheng dianping
952	nonlinear	non-linéaire	nichtlinear	нелинейный	non lineare	sistemas * de navegación	非線形な hi-senkei-na	非线性的 feixianxingde
953	nonlinear control	régulation f non-linéaire	nichtlineare Regelung (Steuerung)	нелинейное управление (регулирование)	controllo non lineare	negación f	非線形制御 hisenki-seigyo	非线性控制 feixianxing kongzhi
954	nonlinear distortion	distorsion f non-linéaire	nichtlineare Verzerrung	нелинейное искажение	distorsione non lineare	realimentación f negativa	非線形歪 hisenki-hizumi	非线性畸变，非线性失真 feixianxing jibian, feixianxing shizhen
955	nonlinear filter	filtre m non-linéaire	nichtlineares Filter	нелинейный фильтр	filtro non lineare	red f	非線形フィルタ hisenki-firuta	非线性滤波器 feixianxing lüboqi
956	nonlinear potentiometer	potentiomètre m non-linéaire	nichtlineares Potentiometer	нелинейный потенциометр	potenziometro non lineare	analizador m de redes	非線形ポテンショメータ hisenki-potensho-mēta	非线性电位计 feixianxing dianweiji
957	nonlinear programing	programmation f non-linéaire	nichtlineare Programmierung	нелинейное программирование	programmazione non lineare	topología * de la red	非線形計画法 hisenki-keikaku-hō	非线性规划 feixianxing guihua
958	nonlinearity	non-linéarité f	Nichtlinearität	нелинейность	non linearità	zona f de insensibilidad, zona f muerta	非線形性 hisenki-sei	非线性 feixianxing
959	nonminimum phase system	système m à déphasage non minimal	nichtminimalphasiges System	неминимально-фазовая система	sistema a fase non minima	control de zona muerta	非最小位相系（システム） hisaishō-isō-kei (-shisutemu)	非最小相位系统 feizuixiao xiangwei xitong
960	nonreturn valve	valve f de retenue	Rückschlagventil, Rückschlagklappe	стопорный клапан, обратный клапан, невозвратный клапан	valvola a ritorno interdetto (unidirezionale)	diagrama m de Nichols	逆止め弁 gyakutome-ben	单向阀 danxiangfa
961	nonstationary system	système m non-stationnaire	nichtstationäres System	нестационарная система	sistema non stazionario	diagrama m de Nichols	非定常系（システム） hiteijō-kei (-shishutemu)	非平稳系统 feipingwen xitong
962	NOR element	circuit m NI, circuit m NON-OU	NOR-Glied	элемент ИЛИ-НЕ	elemento NOR	nodo m	ノア要素 noa-yōso	"或非"元件 "huofei" yuanjian

No.	English	French	German	Russian	Italian	Spanish	Japanese	Chinese
963	NOR operation	opération f NI, opération f NON-OU	NOR-Verknüpfung (-Operation)	операция ИЛИ-НЕ	operazione NOR	operación f NI	ノア演算 noa-enzan	"或非"运算, "或非"操作 "huofei" yunsuan, "huofei" caozuo
964	normal distribution	répartition f normale	Normalverteilung	нормальное распределение, гауссовское распределение	distribuzione normale	distribución f normal, distribución f de Gauss	正規分布 seiki-bunpu	正态分布, 高斯分布 zhengtai fenbu, gaosi fenbu
965	normalized response	réponse f normalisée	normiertes Ansprechverhalten, normierte (System)antwort	нормализованная характеристика	risposta normalizzata	respuesta f normalizada	規準化応答 正規化応答 kijunka-ōtō seikika-ōtō	规范化响应 guifanhua xiangying
966	normally closed contact	contact m de repos	Öffner, Öffnungskontakt, Ruhekontakt	нормально замкнутый контакт	contatto normalmente chiuso	contacto m de reposo	ノーマルクローズ接点 nōmaru-kurōzudo-setten	常闭触点 changbi chudian
967	normally closed gate	porte f normalement fermée	Ausschaltglied, Ausschalttor	нормально закрытый вентиль	passaggio (cancello) normalmente chiuso	puerta f normalmente cerrada	ノーマルクローズドゲート nōmaru-kurōzudo-gēto	常闭门 changbimen
968	normally open contact	contact m de travail	Schließer, Schließkontakt, Arbeitskontakt	нормально разомкнутый контакт	contatto normalmente aperto	contacto m de trabajo	ノーマルオープン接点 nōmaru-ōpun-setten	常开触点 changkai chudian
969	normally open gate	porte f normalement ouverte	Einschaltglied, Einschalter	нормально открытый вентиль	passaggio (cancello) normalmente aperto	puerta f normalmente abierta	ノーマルオープンゲート nōmaru-ōpun-gēto	常开门 changkaimen
970	NOT element	circuit m NON	Nicht-Glied	элемент НЕ, инверторный элемент	negazione o elemento NOT	elemento m NO	否定(ノット)要素 hitei(notto)-yōso	"非"元件 "fei" yuanjian
971	notch filter	filtre m à suppression de bande	Notchfilter, Filter mit Dämpfungspol bei einer Frequenz	фильтр выреза, пазовый фильтр, заграждающий фильтр	filtro a spillo, o a V	filtro * bloqueo	ノッチフィルタ	陷波滤波器 xianbo lüboqi
972	nozzle	tuyère f	Düse, Austrittsdüse	сопло	becco	tobera f	ノズル nozuru	喷嘴 penzui
973	numeric control	régulation f numérique	numerische Regelung (Steuerung)	цифровое управление	controllo numerico	control m numérico	数値制御 sūchi-seigyo	数控 shukong
974	nutation	nutation f	Nutation, Nutationsbewegung	нутация	nutazione	nutación f	章動 shō-dō	章动 zhangdong
975	Nyquist diagram	diagramme m de Nyquist	Nyquistdiagramm	диаграмма Найквиста	diagramma di Nyquist	diagrama m de Nyquist	ナイキスト線図 Nyquist-senzu	奈奎斯特图 naikuisite tu

No.	English	French	German	Russian	Italian	Spanish	Japanese	Chinese
976	Nyquist's criterion	critère m de Nyquist-Cauchy	Nyquistkriterium	критерий Найквиста	criterio di Nyquist	criterio m de Nyquist	ナイキストの (安定) 基準 Nyquist-no-(antei)kijun	奈奎斯特准则 naikuisite zhunze
977	objective function	fonction f téléologique	Zielfunktion, Gütefunktion	функция цели, целевая функция	funzione obiettivo	función f objetivo	目的関数 mokuteki-kansū	目标函数 mubiao hanshu
978	observability	observabilité f	Beobachtbarkeit	наблюдаемость	osservabilità	observabilidad f	可観測性 ka-kansokusei	可观测性 keguancexing
979	observability index	indice m d'observabilité	Beobachtbarkeitsindex	индекс наблюдаемости	indice di osservabilità	índice m de observabilidad	可観測指数 ka-kansoku-shisū	可观测指数 keguance zhishu
980	observable	observable	beobachtbar	наблюдаемый	osservabile	observable	可観測な ka-kansoku-na	可观测的 keguancede
981	observer	observateur m	Beobachter	наблюдатель	osservatore	observador m	観測器 kansoku-ki	观测器 guanceqi
982	offline	hors-ligne, autonome	off-line, leitungs-getrennt, indirekt gekoppelt	офф-лайн, автономный	fuori linea	fuera de línea	オフライン ofurain	离线 lixian
983	offset	statisme m	bleibende Abweichung, statischer Restfehler	постоянное отклонение	deviazione, scostamento	desviación f permanente	オフセット ofusetto	偏移 pianyi
984	online	en ligne, en direct	on-line, leitungsverbunden, direkt gekoppelt	он-лайн	in linea	en línea	オンライン onrain	在线 zaixian
985	online closed loop	boucle f fermée en ligne	on-line Regelkreis	он-лайн замкнутая (система)	anello chiuso in linea	anillo m cerrado conectado en línea	オンライン閉ループ onrain-hei-rūpu	在线闭环回路 zaixian bihuan huilu
986	online control	régulation f en ligne	on-line Regelung (Steuerung)	он-лайн управление	controllo in linea	control m en línea, control m directo	オンライン制御 onrain-seigyo	在线控制 zaixian kongzhi
987	online open loop	boucle f ouverte en ligne	on-line offene Regelschleife, on-line Steuerkette	он-лайн разомкнутая (система), (управление) в режиме советчика	anello aperto in linea	anillo m abierto conectado en línea	オンライン開ループ onrain-kai-rūpu	在线开环回路 zaixian kaihuan huilu
988	on-off action	action f par tout ou rien	Ein-Aus-Verhalten	релейное воздействие	azione a tutto-o-niente	acción f por todo o nada	二動作 オン・オフ動作 nichi-dōsa on-ofu-dōsa	通断作用 tongduan zuoyong

No.	English	French	German	Russian	Italian	Spanish	Japanese	Chinese
989	on-off control	régulation f à action par tout ou rien	Ein-Aus-Regelung, Zweipunktregelung	релейное управление (регулирование)	controllo a tutto-o-niente	control m por todo o nada	オン・オフ制御 二値制御 on-ofu-seigyo nichi-seigyo	通断控制 tongduan kongzhi
990	on-off controller	régulateur m à action par tout ou rien	Ein-Aus-Regler, Zweipunktregler	релейный регулятор	controllore a tutto-o-niente	controlador m todo o nada	オン・オフ調節器 on-ofu-chōsetsuki	通断控制器 tongduan kongzhiqi
991	one-degree-off freedom system	système m à un degré de liberté	System mit einem Freiheitsgrad	система с одной степенью свободы	sistema a un grado di libertà	sistema m con un grado de libertad	1自由度系 ichi-jiyūdo-kei	单自由度系统 danziyoudu xitong
992	open loop	boucle f ouverte	offener Kreis, Steuerkette	разомкнутый контур	anello aperto	anillo m abierto	開回路 kaikairo	开环 kaihuan
993	open-loop control system	système m de commande en boucle	(rückführungsloses) Steuersystem, aufgeschnittenes Reglungssystem	разомкнутая система управления	sistema di controllo ad anello aperto	sistema m de control en anillo abierto	開回路御系 kaikairo-seigyokei	开环控制系统 kaihuan kongzhi xitong
994	open-loop gain	gain m en boucle ouverte	Verstärkung des offenen Regelkreises	коэффициент усиления разомкнутого контура	guadagno d'anello	ganacia f en anillo abierto	開回路ゲイン kaikairo-gein	开环增益 kaihuan zengyi
995	open-loop phase angle	déphasage m en boucle ouverte	Phasenwinkel des offenen Kreises	фазовый сдвиг разомкнутого контура	sfasamento ad anello aperto	desfase m en anillo abierto	開回路位相角 kaikairo-isōkaku	开环相角 kaihuan xiangjiao
996	open-loop transfer function	transmittance f en boucle ouverte	Übertragungsfunktion des offenen Kreises	передаточная функция разомкнутой системы	funzione di trasferimento d'anello	función f de transferencia en anillo abierto	開回路伝達関数 kaikairo-dentatsu-kansū	开环传递函数 kaihuan chandi hanshu
997	operate (v)	agir, actionner (v)	betreiben, betätigen, ablaufen	приводить в действие; воздействовать, возбуждать	(far) funzionare	funcionar, operar	操作する sōsa-suru	操作、运算 (动词) caozuo, yunsuan (dongci)
998	operating range	domaine m d'action	Arbeitsbereich	рабочий диапазон	campo di funzionamento	campo m de funcionamiento	動作範囲 dōsa-han-i	运算范围、操作范围 yunsuan fanwei, caozuo fanwei
999	operating system	système m d'exploitation	Betriebssystem	операционная система	sistema operativo	sistema m operativo	オペレーティングシステム kanri-shisutemu	操作系统 caozuo xitong
1000	operational amplifier	amplificateur m opérationnel	Operationsverstärker	операционный усилитель	amplificatore operazionale	amplificador m operacional	演算増幅器 enzan-zōfukuki	运算放大器 yunsuan fangdaqi
1001	operational calculus	calcul m opérationnel	Operatorenrechnung	операционное исчисление	calcolo operazionale	cálculo m operacional	演算子法 enzan	运算微分 yunsuan weijifen

No.	English	French	German	Russian	Italian	Spanish	Japanese	Chinese
1002	optical character recognition	reconnaissance f optique de caractères	optische Zeichenerkennung	оптическое распознавание символов	riconoscimento ottico dei caratteri	reconocimiento óptico de caracteres	光学文字認識 kōgaku-moji-ninshiki	光学特性识别 guangxue texing shibie
1003	optimal control	commande f optimale	optimale Regelung (Steuerung)	оптимальное управление	controllo ottimo	control m óptimo	最適制御 saiteki-seigyo	最优控制 zuiyou kongzhi
1004	optimal estimation	estimation f optimale	optimale Schätzung	оптимальное оценивание	stima ottima	estimación f óptima	最適推定 saiteki-suitei	最优估计 zuiyou guji
1005	optimal system	système m optimal	optimales System	оптимальная система	sistema ottimale	sistema m óptimo	最適系（システム） saiteki-kei(-shisutemu)	最优系统 zuiyou xitong
1006	optimal trajectory	trajectoire f optimale	optimale Trajektorie	оптимальная траектория	traiettoria ottima	trayectoria * óptima	最適軌道 saiteki-kidō	最优轨迹 zuiyou guiji
1007	optimization	optimisation f	Optimierung	оптимизация	ottimizzazione	optimización f	最適化 saitekika	优化 youhua
1008	optimize (v)	optimiser (v)	optimieren	оптимизировать	ottimizzare	optimizar	最適化する saitekika-suru	使优化 (动词) shiyouhua (dongci)
1009	OR element	circuit m OU	ODER-Glied (Gatter)	элемент ИЛИ	elemento OR	elemento m O	論理和要素 ronriwa-yōso	"或"元件 "huo" yuanjian
1010	OR operation	opération f OU	Disjunktion, ODER-Verknüpfung (od. Operation)	операция ИЛИ	operazione OR	operación f O	論理和演算 ronriwa-enzan	"或"运算，"或"操作 "huo" yunsuan, "huo" caozuo
1011	oscillation	oscillation f	Oszillation, Schwingung	колебания	oscillazione	oscilación f	振動 shindō	振荡 zhendang
1012	outer gimbal	Cardan m extérieur	Kardanrahmen	рама кардана	sospensione cardanica esterna	suspensión f cardánica exterior	外部ジンバル gaibu-jinbaru	外框 waikuang
1013	outer loop	boucle f principale	äußere Schleife, äußerer Kreis	основной контур	anello esterno	anillo m externo, anillo m principal	外部ループ gaibu-rūpu	外回路 waihuilu
1014	output	grandeur f de sortie	Ausgangsgröße	выход	uscita	salida f	出力 shutsuryoku	输出 shuchu

	English	French	German	Russian	Italian	Spanish	Japanese	Chinese
1015	output axis	axe *m* de sortie	Abgriffsachse	выходная ось	asse d'uscita	eje de salida	出力軸 shutsuryoku-jiku	输出轴 shuchuzhou
1016	output brushes	balais *m* principaux	Ausgangsbürsten	выходные щетки	spazzole d'uscita	escobillad de salida	出力ブラシ shutsuryoku-burashi	输出电刷 shuchu dianshua
1017	output-decentralized system	système *m* à grandeurs de sortie décentralisées	ausgangsdezent-ralisiertes System	система с децентрализованным выходом	sistema decentralizzato in uscita	sistema *m* de salida descentralizada	出力分散システム shutsuryoku-bunsan shisutemu	分散输出系统 fensan shuchu xitong
1018	output matrix	matrice *f* de sortie	Ausgangsmatrix	матрица выхода	matrice d'uscita	matriz *f* de salida	出力行列 shutsuryoku-gyōretsu	输出矩阵 shuchu juzhen
1019	output signal	signal *m* de sortie	Ausgangssignal	выходной сигнал, выходное воздействие	segnale d'uscita	señal *f* de salida	出力信号 shutsuryoku-shingō	输出信号 shuchu xinhao
1020	output transfer function	transmittance *f* en boucle fermée	Ausgangsübertragungs-funktion	передаточная функция замкнутой системы	funzione di trasferimento all'uscita	función *f* de transferencia en anillo cerrado	出力伝達関数 shutsuryoku-dentatukansū	输出传递函数 shuchu chuandi hanshu
1021	output variable	variable *f* de sortie	Ausgangsvariable	выходная величина	variabile d'uscita	variable *f* de salida	出力変数 shutsuryoku-hensū	输出变量 shuchu bainliang
1022	output winding	enroulement *m* de sortie	Lesewicklung	выходная обмотка	avvolgimento d'uscita	arrollamiento *m* de salida	出力巻線 shutsuryoku-makisen	输出绕组 shuchu raozu
1023	overdamping	sur-amortissement *m*	aperiodische Dämpfung	апериодическое демпфирование	sovrasmorzamento	amortiguamiento *m* supercrítico	過減衰 ka-gensui	过阻尼 guozuni
1024	overlap	recouvrement *m*, empiétement *m*	Überdeckung, Übergreifen	перекрытие, наложение	sovrapposizione	solape *m*	重合 jūgō	重叠 chongdie
1025	overlapping	recouvrement *m*	Überlappung, überlappende Befehlsausführung od. -Verwaltung	перекрытие, наложение	sovrapposizione	recubrimiento *m*, solapamiento *m*	重合 jūgō	重叠 chongdie
1026	overload	surcharge *f*	Überlastung, Mehrbelastung	перегрузка	svraccarico	sobrecarga *f*	過負荷 kafuka	过载 guozai
1027	overshoot	dépassement *m*, taux *m* de dépassement	Überschwingung, übersteuern	перерегулирование	sovrelongazione	rebasamiento *m*, sobrepaso *m*	行き過ぎ量 yukisugi-ryō	超调 chaotiao

	English	French	German	Russian	Italian	Spanish	Japanese	Chinese
1028	page printer	imprimante f (à pages)	Blattschreiber, Seitendrucker	страничный принтер	stampante a pagina	impresora f de páginas	ページプリンタ peiji-purinta	页式打印机 yeshi dayinji
1029	parallel processing	traitement m en parallèle	Parallelverarbeitung	параллельная обработка	elaborazione parallela	procesamiento * paralelo	並列処理 heiretsu-shori	并行处理 bingxing chuli
1030	parallel-sequential method	méthode f parallèle-sérielle	paralle-serielle Übertragung	параллельно-последовательный метод	metodo parallelo-sequenziale	método m secuencial en paralelo		并行时序方法 bingxing shixu fangfa
1031	parallel transducer	transducteur m parallèle	Transduktor in Parallelschaltung	параллельный магнитный усилитель	trasduttore in parallelo	transductor m paralelo	並列トランスダクタ heiretsu-toransudakuta	并行饱和电抗器 bingxing baohe diankangqi
1032	parameter	paramètre m	Parameter	параметр	parametro	parámetro m	パラメータ paraméta	参量, 参数 canliang, canshu
1033	parameter estimation	estimation f du paramètre	Parameterschätzung	оценивание параметров	stima dei parametri	estimación * de parámetros	パラメータ推定 paraméta-suitei	参数估计 canshu guji
1034	parametric variation	variation f paramétrique	parametrische Variation (od. Änderung)	параметрическая вариация	variazione parametrica	varición f paramétrica	パラメータ変動 paraméta-hendō	参数变化 canshu bianhua
1035	parity check	contrôle m de parité	Paritätsprüfung	проверка на чётность	verifica di parità	comprobación * de paridad	奇遇検査 kigū-kensa pariti-kensa	奇偶校验 ji'ou jiaoyan
1036	partial connective stability	stabilité f à connexion partielle	partielle Verbundstabilität, Teilverbundstabilität	частично связующая стабильность	stabilità connettiva parziale	estabilidad conectiva parcial	連結部安定 tsūka-tai-iki	部分连接稳定性 bufen lianjie wendingxing
1037	pass-band	passe-bande m	Durchlässigkeitsbereich	полоса пропускания	passa-banda	banda de paso, banda f pasante	通過帯域 tsūka-tai-iki	通频带 tongpindai
1038	passive compensation	compensation f passive	passive Kompensation	пассивная компенсация	compensazione passiva	compensación pasiva	受動的補償 judōteki-hoshō	无源补偿 wuyuan buchang
1039	passive element	organe m passif	passives Element	пассивный элемент	elemento passivo	elemento m pasivo	受動要素 judō-yōso	无源元件 wuyuan yuanjian
1040	passive filter	filtre m passif	passives Filter	пассивный фильтр	filtro passivo	filtro * pasivo	受動フィルタ judō-firuta	无源滤波器 wuyuan lüboqi

No.	English	French	German	Russian	Italian	Spanish	Japanese	Chinese
1041	path	voie *f*	Pfad, Kanal, Laufband	путь	cammino, percorso	camino *m*	径路 keiro	路径 lùjìng
1042	pattern generation	génération *f* de formes	Mustergenerierung (od. Erzeugung)	генерирование образа	generazione di configurazioni	generación *f* de formas, generación *f* de patrones	パターン生成 patān-seisei	模式发生 moshi fasheng
1043	pattern identification	identification *f* de formes	Zeich- od. Musteridentifikation	распознавание образов	identificazione di configurazioni	identificación *f* de formas	パターン同定 patān-dōtei	模式辨识 moshi bianshi
1044	pattern recognition	reconnaissance *f* de(s) formes	Zeichen od. Mustererkennung	распознавание образов	riconoscimento di configurazioni	reconocimiento *m* de formas	パターン認識 patān-ninshiki	模式识别 moshi shibie
1045	pay-off function	fonction *f* bénéfice	Gewinnfunktion, Betragsfunktion	платежная функция	funzione d'introito	función *f* de coste	支付函数 zhifu hanshu	
1046	performance characteristic	caractéristique *f* d'efficacité	Leistungskennwert, Leistungsgröße, Leistungsmerkmal	характеристика качества	caratteristica di merito	características *f* de comportamiento, características *f* funcionales	評価特性 hyōka-tokusei	性能特征 xingneng tezheng
1047	performance function	fonction *f* d'efficacité	Zielfunktion, Gütefunktion	функция качества	funzione di merito	función *f* de comportamiento	評価関数 hyōka-kansū	性能函数 xingneng hanshu
1048	performance index	indice *m* d'efficacité	Zielgröße, Gütegröße	критерий качества	indice di prestazione, o cifra di merito	índice *m* de comportamiento índice *m* e funcionamiento	評価指数 hyōka-shisū	性能指标 xingneng zhibiao
1049	periodic wave	onde *f* périodique	periodische Welle	периодический сигнал	onda periodica	onda *f* periódica	周期波 shūki-ha	周期波 zhouqibo
1050	permanent-magnet motor	motormoteur *m* à aimant permanent	Permanentmagnetmotor	мотор с постоянным магнитом	motore a magnete permanente	motor *m* de imán permanente	永久磁石電動機 eikyū-jishaku-dendōki	永磁电机 yongci dianji
1051	perturbation	perturbation *f*	Störung	возмущение	perturbazione	perturbación *f*	パーターベーション 摂動 pātabeishon sestudō	摄动, 微扰 shedong, weirao
1052	perturbation theory	théorie *f* des perturbations	Störungstheorie	теория возмущений	teoria delle perturbazioni	teoría *f* de perturbaciones	摂動理論 setsudō-riron	摄动理论 shedong lilun
1053	phase	phase *m*	Phase	фаза	fase	fase *f*	位相 isō	相, 相位 xiang, xiangwei

	English	French	German	Russian	Italian	Spanish	Japanese	Chinese
1054	phase advance	avance f de phase	Phasenvoreilung, Phasenvorlauf	опережение по фазе	anticipo di fase	advance m de fase	位相進み isō-susumi	相位超前 xiangwei chaoqian
1055	phase-advance network	réseau m d'avance de phase	phasenvoreilendes Netzwerk	цепь опережения	rete anticipatrice	red de avance de fase	位相進み回路 isō-susumi-kairo	相位超前网络 xiangwei chaoqian wangluo
1056	phase angle	angle m de phase	Phasenwinkel	фазовый угол	angolo di fase	ángulo m de fase	位相角 isō-kaku	相角，相位移 xiangjiao, xiangweiyi
1057	phase characteristic	caractéristique f de phase	Phasenverlauf	фазовые характеристики	caratteristica di fase	característica f de fase	位相特性 isō-tokusei	相位特性 xiangwei texing
1058	phase contour	courbe f de déphasage	Phasenkontur	равнофазный контур	linea di livello della fase	curva f de fase	位相軌跡 isō-kiseki	等相线 dengxiangxian
1059	phase crossover frequency	fréquence f de coupure de phase	Phasendurchtritts-frequenz	частота среза по фазе	frequenza (o pulsazione) di taglio (nel diagramma) delle fasi	frecuencia f de cruce de fase	位相交点周波数 isōkōten-shūhasū	相位穿越频率 xiangwei chuanyue pinlü
1060	phase difference	différence f de phase	Phasendifferenz, Phasenunterschied, Gangunterschied	разность фаз	sfasamento	diferencia f de fase	位相差 isō-sa	相位差 xiangweicha
1061	phase distortion	distorsion f de phase	Phasenverzerrung, Phasendrehung	фазовое искажение	distorsione di fase	distorsión f de fase	位相歪 isō-hizumi	相位畸变，相位失真 xiangwei jibian, xiangwei shizhen
1062	phase inverter	amplificateur m inverseur	Phasenumkehrstufe, Phaseninverter	фазоинвертер	invertitore di fase	inversor m de fase	位相反転器 isō-hantenki	倒相器 daoxiangqi
1063	phase lag	retard m de phase	Phasennacheilung, Phasenverzögerung	отставание по фазе	ritardo di fase	retardo m de fase	位相おくれ isō-okure	相位滞后 xiangwei zhihou
1064	phase lead	avance f de phase	Phasenvoreilung	опережение по фазе	anticipo di fase	avance m de fase	位相すすみ isō-susumi	相位超前 xiangwei chaoqian
1065	phase-locked loop	boucle f à verrouillage de phase, d'asservissement de phase	phasenstarre Regelschleife, phasenstarrer Regelkreis	контур фазовой автоподстройки	anello ad aggancio di fase	bucle de encaje de fase	フェーズロックループ、PLL feizurokku-rūpu	锁相回路 suoxiang huilu
1066	phase locus	lieu m de phase	Phasenortskurve	фазовый годограф	diagramma delle fasi	lugar m (geométrico) de fases	位相軌跡 isō-kiseki	相轨迹 xiangguiji

No.	English	French	German	Russian	Italian	Spanish	Japanese	Chinese
1067	phase margin	marge f de phase	Phasenrand	запас по фазе, избыток фазы	margine di fase	margen m de fase	位相余有 isō-yoyū	相位裕度, 相位裕量 xiangwei yudu, xiangwei yuliang
1068	phase plane	plan m de phase	Phasenebene	фаовая плоскость	piano di fase	plano m de fases	位相面 isō-men	相平面 xiangpingmian
1069	phase response	réponse f en phase	Phasenantwort, Phasengang	фазовая характеристика	risposta in fase	respuesta f de fase	位相応答 isō-ōtō	相位响应 xiangwei xiangying
1070	phase shift	déphasage m	Phasenverschiebung	сдвиг фазы	variazione di fase	desfase m, desfasaje m	位相ずれ isō-zure	相移 xiangyi
1071	phase space	espace m de phase	Phasenraum	фазовое пространство	spazio delle fasi	espacio m de fases	位相空間 isō-kūkan	相空间 xiangkongjian
1072	photocell pick-off	capteur m photo électrique	Photozellenabgriff	фотоэлектрический датчик	captatore a fotocellula	selección por fotocélula	光電検出 kōden-kenshutsu	光电管检出 guangdian jianchu
1073	pick-off	capteur m	Abgriff	датчик	captatore	selección	引き出し hiki-dashi	检出 jianchu
1074	picture element	circuit m graphique	Bildelement, Bildpunkt	элемент рисунка	elemento di figura	elemento * gráfico	画像要素 gazō-yōso	象素, 象元 xiangsu, xiangyuan
1075	piecewise linear	linéaire par segment, par section	stückweise linear	кусочно линейный	lineare a tratti	lineal a trozos	断片線形 danpen-senkei	分段线性 (的) fenduan xianxing (de)
1076	pilot (v)	piloter (v)	führen, leiten, lenken, lotsen, steuern	стабилизировать, управлять	pilotare	pilotar	案内する annai-suru	导引 (动词) daoyin (dongci)
1077	pilot operation	opération f de pilotage	Steuervorgang	стабилизация, вспомогательное управление	operazione pilota	pilotaje m	パイロット操作 pairotto-sōsa	导引操作 daoyin caozuo
1078	pilot valve	vanne f pilote	Schaltventil, Steuerventil	вспомогательный клапан	valvola pilota	válvula f piloto	パイロット弁 pairotto-ben	导引阀, 控制阀 daoyinfa, kongzhifa
1079	pinwheel gear	roue f à chevilles, roue f à taquets	Stiftradgetriebe	цевочное зубчатое колесо	ingranaggio elicoidale	engranaje de rueda de aguja	ピン歯車 pin-haguruma	针轮机构 zhenlun jigou

No.	English	French	German	Russian	Italian	Spanish	Japanese	Chinese
1080	pipelining processing	traitement *m* en série	Pipelineverarbeitung	конвейерная обработка	elaborazione a convogliamento di flusso	procesamiento * segmentado	パイプライン処理 paipurain-shori	流水线处理 liushuixian chuli
1081	piston valve	vanne *f* à piston	Kolbenventil	поршневой клапан	valvola a pistone	válvula *f* de pistón	ピストン弁 pisuton-ben	滑阀 huafa
1082	pixel	élément *m* d'image, point *m*	Bildelement, Pixel	пиксель	'pixel'	píxel	ピクセル	象素，象元 xiangsu, xiangyuan
1083	plant	système *m* industriel	Regelstrecke, Anlage, Prozeßanlage	объект (управления)	impianto	instalación *f*, planta *f*	プラント puranto	对象 duixiang
1084	play	jeu *m*	Spiel, Lose	люфт	gioco	juego *m*	遊び asobi	动作，游戏 dongzuo, youxi
1085	pneumatic amplifier	amplificateur *m* pneumatique	pneumatischer Verstärker	пневмоусилитель	amplificatore pneumatico	amplificador *m* neumático	空気式増幅器 kūkishiki-zōfukuki	气动放大器 qidong fangdaqi
1086	pneumatic relay	relais *m* pneumatique	pneumatisches Relais	пневмореле	commutatore pneumatico	relé *m* neumático	空気継電器 kūkiatsu-rirē	气动继电器 qidong jidianqi
1087	point of control	emplacement *m* d'action du réglage	Stellort	место приложения исполнительного воздействия	punto di controllo	punto *m* de funcionamiento	制御点 seigyoten	控制点 kongzhidian
1088	polar plot	représentation *f* polaire	Frequenzgang-darstellung (in komplexer Zahlenebene)	полярная диаграмма	diagramma polare	gráfico *m* polar	極座標プロット kyokuzahyō-purotto	极座标图 jizuobiaotu
1089	pole	pôle *m*	Pol	полюс	polo	polo *m*	極 kyoku	极点 jidian
1090	pole assignment	placement *m* de pôles	Polzuweisung, Polfestlegung	размещение полюсов	assegnamento dei poli	asignación *f* de polos	極配置 kyokuhaichi	极点配置 jidian peizhi
1091	pole-zero assignment	placement *m* des pôles et zéros	Pol-Nullstellenfestlegung	размещение полюсов и нулей	assegnamento di poli e zeri	asignación *f* de polos y ceros	極一零(点)配置 kyoku-zero(ten)-haichi	零极点配置 lingjidian peizhi
1092	Popov criterion	critère *m* (de stabilité absolue)	Popov-Kriterium	критерий Попова	criterio di Popov	criterio * de Popov	ポポフの条件 Popov-no-jyōken	波波夫准则 bobofu zhunze

	English	French	German	Russian	Italian	Spanish	Japanese	Chinese
1093	- position	position f	Position	положение	posizione	posición f	位置 ichi	位置 weizhi
1094	position control	réglage m, asservissement m de position	Positionsregelung	управление положением	controllo di posizione	control de posición	位置制御 ichi-seigyo	位置控制 weizhi kongzhi
1095	position error	erreur f de position	Positionsfehler	позиционная погрешность	errore di posizione	error m de posición	位置偏差 ichi-hensa	位置误差 weizhi wucha
1096	position feedback	réaction f de mise en position	Positionsrückführung	обратная связь по положению	retroazione di posizione	realimentación f de posición	位置のフィードバック ichi-no-fidobakku	位置反馈 weizhi fankui
1097	positive-displacement pump	pompe f volumétrique	volumetrische Pumpe, Pumpe in Verdrängungsbauart	поршневой насос	pompa a dislocamento positivo	bomba f volumétrica	容積式ポンプ yōsekishiki-ponpu	正排量泵 zhengpailiangbeng
1098	positive feedback	réaction f positive	positive Rückführung	положительная обратная связь	retroazione positiva	realimentación f positiva	正のフィードバック sei-no-fidobakku	正反馈 zhengfankui
1099	positive-negative three-level action	action f par plus ou moins	dreiwertige Steuergröße (plus-minus-Null)	трехпозиционное воздействие	azione a tre livelli positivo-negativa	acción de tres niveles positiva-negativa	正-負 3位置動作 sei-fu 3ichi-dōsa	正负三位作用 zhengfu sanwei zuoyong
1100	potential divider	diviseur m de tension	Spannungsteiler	делитель напряжения	partitore di tensione	divisor m de tensión	分圧器 bun-atsuki	分压器 fenyaqi
1101	potentiometer	potentiomètre m	Potentiometer	потенциометр	potenziometro	potenciómetro m	ポテンショメータ den-i-sakei	电位器 dianweiqi
1102	potentiometer pick-off	capteur m potentiométrique	Potentiometerabgriff	потенциометрический датчик	captatore potenziometrico	selección por potenciómetro	ポテンショメータ検出 potensho-mēta-kenshutsu	电位器检出 dianweiqi jianchu
1103	power amplifier	amplificateur m de puissance	Leistungsverstärker	усилитель мощности	amplificatore di potenza	amplificador m de potencia	電力増幅器 denryoku-zōfukuki	功率放大器 gonglü fangdaqi
1104	power-assisted control (closed loop)	régulation f indirecte, servocommande f	Regelung mit Hilfsenergie, Servoregelung	непрямое управление (регулирование) (замкнутое)	controllo (ad anello chiuso) ad alimentazione assistita	servocontrol m	他力制御 tariki-seigyo	动力辅助控制 (闭系) dongli fuzhu kongzhi (bihuan)
1105	power circuit	circuit m de puissance	Arbeitsstromkreis	силовая цепь	circuito di potenza	circuito m de potencia	電源回路 dengen-kairo	电源电路 dianyuan dianlu

No.	English	French	German	Russian	Italian	Spanish	Japanese	Chinese
1106	power density spectrum	puissance f de densité spectrale	spektrale Leistungsdichte,	спектральное распределение плотности (энергии)		espectro m de densidad de potencia	パワースペクトル密度 pāwa-supekutoru-mitsudo	功率谱 gonglüpu
1107	power winding	enroulement m de puissance	Leistungswicklung	силовая обмотка		arrollamiento m de potencia, devanado m de potencia	電力巻線 dennyoku-makisen	功率绕组 gonglü raozu
1108	precess (v)	précéder (v)	präzedieren	предварять	precedere	precesionar (efectuarse la precesión)	歳差運動する saisa-undō-suru	进动，旋动 (动词) jindong, xuandong (dongci)
1109	precision	précision f	Genauigkeit(sgrad), Meßgenauigkeit	точность	precisione	precisión f	精度 seido	精（密）度 jing(mi)du
1110	prediction error method	méthode f de l'erreur de prédiction	Fehlervorhersageverfahren, Fehlerprädiktionsverfahren	метод предсказуемой ошибки	metodo a predizione dell'errore	método m de la predicción del error	予測誤差法 yosokugosa-hō	预报误差方法 yubao wucha fangfa
1111	predictor theory	théorie f du prédicteur, de la prévision	Prädiktionstheorie, Vorhersagetheorie	теория прогнозирования	teoria dei predittori	teoría * de predictores	予測理論 yosokuki-riron	预报理论 yubao lilun
1112	preprocessing	prétraitement m	(Daten-) Vorverarbeitung	предварительная обработка	pre-elaborazione	preprocesamiento m	前処理 maeshori	预处理 yuchuli
1113	preprocessor	préprocesseur m	Vorlaufprogramm, Vorrechner	препроцессор	pre-elaboratore	preprocesador m	前処理プログラム maeshori-puroguramu	预处理器 yuchuliqi
1114	preset (v)	afficher par avance (v)	vorher festgelegt od. festlegen, vorwählen	(предварительно) устанавливать	prefissare	preajustar, prefijar	プレセット頂数する	预置，预调 (动词) yuzhi, yutiao (dongci)
1115	primary detecting element	capteur m	Aufnehmer, Meßfühler (erstes Element eines Meßkreises)	детектор	elemento rivelatore primario	captador m, detector m	主検出要素 shu-kensyutsu-yōso	初级检测元件 chuji jiance yuanjian
1116	primary regulation	régulation f primaire	Primäreinstellung, Primärregulation	первичное регулирование	regolazione primaria	regulación f primaria	一次調節 ichiji-chōsetsu	一次调节 yici tiaojie
1117	principle of superposition	principe m de superposition	Superpositionsprinzip	принцип суперпозиции	principio di sovrapposizione	principio m de superposición	重ね合わせの原理 kasane-awase-no-genri	叠加原理 diejia yuanli
1118	printed circuit	circuit m imprimé	gedruckte Schaltung, gedruckter Schaltkreis	печатная схема	circuito stampato	circuito m impreso	プリント回路 purinto-kairo	印刷电路 yinshua dianlu

No.	English	French	German	Russian	Italian	Spanish	Japanese	Chinese
1119	printer	imprimante f	Drucker	печатающее устройство, принтер	stampante	impresora f	プリンタ purinta	打印机 dayinji
1120	probabilistic logic	logique f stochastique	Wahrscheinlichkeitslogik	вероятностная логика	logica probabilistica	lógica f probabilística	確率論理 kakuritsu-ronri	概率性逻辑 gailüxing luoji
1121	probabilistic model	modèle m stochastique	Wahrscheinlichkeitsmodell	вероятностная модель	modello probabilistico	modelo m probabilístico	確率モデル kakuritsu-moderu	概率模型 gailü moxing
1122	probability	probabilité f	Wahrscheinlichkeit	вероятность	probabilità	probabilidad f	確率 kakuritsu	概率 gailü
1123	probability density function	fonction f de densité de probabilité	Wahrscheinlichkeitsdichte(-funktion)	функция плотности вероятности	funzione densità di probabilità	función f de densidad de probabilidad	確率密度関数 kakuritsu-mitsudo-kansū	概率密度函数 gailü midu hanshu
1124	probability distribution function	fonction f de répartition de probabilité	Wahrscheinlichkeitsverteilung	функция распределения вероятности	funzione distribuzione di probabilità	función f de distribución de probabilidad	確率分布関数 kakuritu-bunpu-kansū	概率分布函数 gailü fenbu hanshu
1125	problem-oriented language	langage m orienté vers la résolution de problèmes	problemorientierte Programmiersprache	проблемно-ориентированный язык	linguaggio orientato al problema	lenguaje m orientado hacia el problema	問題指向型言語 mondai-shikōgata-gengo	面向问题的语言 mianxiang wenti de yuyan
1126	procedure-oriented language	langage m orienté vers les méthodes procédurales	verfahrensorientierte Programmiersprache	процедурно-ориентированный язык	linguaggio orientato alle procedure	lenguaje m orientado hacia el procedimiento	手続指向型言語 tetsuzuki-shikōgata-gengo	面向过程的语言 mixiang guochengde yuyian
1127	process	procédé m	Prozeß, Vorgang, Durchführung, Bearbeitung	(производственный) процесс	processo	proceso m	プロセス purosesu	过程 guocheng
1128	process automation	automatisation f de procédés	Prozeßautomatisierung	автоматизация процесса	automazione di processo	automatización f de procesos	プロセスオートメーション purosesu-ōtomeishon	过程自动化 guocheng zidonghua
1129	process computer	calculateur m de procédés	Prozeßrechner	УВМ, управляющая ЭВМ	calcolatore di processo	computador m de procesos	プロセス計算機 purosesu-keisanki	过程计算机 guocheng jisuanji
1130	process control (closed loop)	régulation f d'un processus	Prozeßregelung (geschlossener Kreis)	управление процессом в замкнутой системе	controllo dei processi (ad anello chiuso)	control m de procesos (anillo cerrado)	プロセス制御 (閉回路) purosesu-seigyo (heikairo)	过程控制 (闭环) guocheng kongzhi (bihuan)
1131	process control (general term)	conduite f d'un processus, commande f d'un processus	Prozeßregelung Prozeßsteuerung (Sammelbegriff)	управление (производственным) процессом (общий термин)	controllo dei processi (termine generale)	control m de procesos (término general)	プロセス制御 (一般用語) purosesu-seigyo (ippan-yōgo)	过程控制 (一般词) guocheng kongzhi (yibanci)

No.	English	French	German	Russian	Italian	Spanish	Japanese	Chinese
1132	process control (open loop)	commande f directe d'un processus	Prozeßsteuerung (offener Kreis)	управление процессом в разомкнутой системе	controllo dei processi (ad anello aperto)	control m de procesos (anillo abierto)	プロセス制御 (開回路) purosesu-seigyo (kaikairo)	过程控制 (开环) guocheng kongzhi (kaihuan)
1133	process control language	langage m de régulation de procédés	(höhere) Prozeßsteuerungssprache	язык управления процессом	linguaggio per il controllo dei processi	lenguajes m de control de procesos	プロセス制御言語 purosesu-seigyo-gengo	过程控制语言 guocheng kongzhi yuyan
1134	process equipment	installation f	Prozeßeinrichtung, Prozeßanlage, Prozeßgeräte	объект (управления)	equipaggiamento di processo	equipo m, instalación f	プロセス装置 purosesu-sōchi	生产设备, 过程装置 shengchan shebei, guocheng zhuangzhi
1135	process identification	identification f des procédés	Prozeßidentifikation	идентификация процесса	identificazione del processo	identificación f de procesos	プロセス同定 purosesu-dōtei	过程辨识 guocheng bianshi
1136	process model	modèle m de procédé	Prozeßmodell	модель процесса	modello del processo	modelo m del proceso	プロセスモデル purosesu-moderu	过程模型 guocheng moxing
1137	processor	processeur m	Prozessor	процессор	processore, elaboratore	procesador m	処理装置 purosessa	处理器 chuliqi
1138	program	programme m	Programm	программа	programma	programa m	プログラム puroguramu	程序 chengxu
1139	program assembler	programme m assembleur	Programmassembler	ассемблер-программа	assemblatore	ensamblador *	プログラムアセンブラ puroguramu-asenbura	汇编语言 huibian yuyan
1140	program controlling element (US) programmer	programmateur m	Programmgeber	программный элемент	elemento di controllo del programma	(elemento) programador m	プログラム制御要素・プログラマー puroguramu-seigyo-yōso, puroguramu	程控装置编程器 chengkong zhuangzhi bianchengqi
1141	programable read only memory (PROM)	mémoire f morte programmable	programmierbarer Festwertspeicher (PROM)	полупостоянная память, программируемая постоянная память	memoria programmabile a sola lettura (PROM)	memoria * programable de sólo lectura	プログラム可能な固定記憶装置 puroguramu-kanōna-koteikiokusōchi	可编程只读存储器 kebiancheng zhidu cunchuqi
1142	program controller	régulateur m à programme	Programmregler, Programmschaltgerät	программный регулятор	controllore di programma	controlador m programable	プログラム調節計 puroguramu-chōsetsu-kei	程序控制器 chengxu kongzhiqi
1143	program controlling element	programmateur m	Programmschaltwerk	программный элемент	elemento di controllo del programma	(elemento) programador m	プログラム調節要素 puroguramu-chōsetsu-yōso	程序控制装置 chengxu kongzhi zhuangzhi
1144	program store	mémoire f à programme	Programmspeicher	накопитель программы	(memoria dedicata ad) archivio programmi	memoria f de programa	プログラム記憶 puroguramu-kioku	程序存储器, 程序存储 chengxu cunchuqi, chengxu cunchu

No.	English	Français	Deutsch	Русский	Italiano	Español	日本語	中文
1145	programable controller	régulateur m programmable	programmierbarer Regler	программируемый контроллер	controllore programmabile	controlador m programable	プログラマブル制御装置 puraguramaburu-seigyo-sōchi	可编程序控制器 kebian chengxu kongzhiqi
1146	programable logic (PLC) controller	automate m programmable	programmierbare Verknüpfungssteuerung (PLC)	программируемый логический контроллер	controllore a logica programmabile (PLC)	autómata m programable, controlador m lógico programable	プログラマブルシーケンス制御装置 puroguramaburu-shīkensu-seigyo-sōchi	可编程序逻辑控制器 kebian chengxu luoji kongzhiqi
1147	programed control	commande f à programme	programmierte Regelung, Progammsteuerung	программное управление (регулирование)	controllo programmato	control m por programa	プログラム制御 puroguramu-seigyo	程序控制 chengxu kongzhi
1148	programing language	langage m de programmation	Programmiersprache	язык программирования	linguaggio di programmazione	lenguaje m de programación	プログラム言語 puroguramu-gengo	编程语言 biancheng yuyan
1149	programing system	système m de programmation	Programmsystem	система программирования	sistema di programmazione	sistema m de programación	プログラミングシステム puroguramingu-shisutemu	编程系统 biancheng xitong
1150	progressive action	action f progressive	progressives Verhalten	непрерывное воздействие	azione progressiva	acción f progresiva	逐次動作 chikuji-dōsa	累进作用 leijin zuoyong
1151	proportional action	action f proportionnelle, action f P	Proportionalverhalten, P-Verhalten	воздействие по отклонению	azione proporzionale	acción f proporcional, acción f proporcional	比例動作 hirei-dōsa	比例作用 bili zuoyong
1152	proportional action factor	facteur m d'action proportionnelle	Proportionalbeiwert	коэффициент воздействия по отклонению	coefficiente dell'azione proporzionale	coeficiente m de acción proporcional	比例動作係数 hirei-dōsa-keisū	比例作用因子 bili zuoyong yinzi
1153	proportional band	étendue f proportionnelle	Proportionalbereich	зона пропорционального регулирования	banda proporzionale	banda f proporcional, margen m proporcional	比例帯 hirei-tai	比例带 bilidai
1154	proportional control factor	gain m d'un régulateur proportionnel	proportionaler Regelfaktor, Proportionalaussteuerung	коэффициент управления по отклонению	coefficiente dell'azione (o del controllo) proporzionale	coeficiente m de control proporcional ganancia f de un regulador proporcional	比例制御係数 hirei-seigyo-keisū	比例控制系数 bili kongzhi xishu
1155	proportional controller	régulateur m proportionnel	Proportionalregler, P-Regler	П-регулятор, пропорциональный регулятор	controllore proporzionale	regulador m de acción proporcional	比例調節器 hireichōsetsuki	比例控制器 bili kongzhiqi
1156	proportional controller with disturbance-variable compensation	régulateur m proportionnel avec compensation des perturbation	P-Regler mit Störkompensation	пропорциональный регулятор с компенсацией возмущения	controllore proporzionale con compensazione del disturbo	regulador m proporcional con compensación, de la variable de perturbación	外乱補償比例調節器 gairan-hoshō-hirei-chōsetsuki	扰动补偿比例控制器 raodong buchang bili kongzhiqi
1157	proportional gain	gain m proportionnel	Proportional-verstärkung	пропорциональное усиление	guadagno proporzionale	ganancia f proporcional	比例ゲイン hirei-gein	比例增益 bili zengyi

№	English	français	Deutsch	Русский	italiano	español	日本語	中文
1158	proportional-plus-derivative controller	régulateur m proportionnel et par dérivation	PD-Regler, Regler mit proportionalem u. differenzierendem Verhalten	ПИД-регулятор, пропорционально-дифференциальный регулятор	controllore proporzionale-derivativo	regulador m de acción proporcional y derivada	比例＋微分調整器 hirei+bibun-chōsetsuki	比例微分控制器 bili weifen kongzhiqi
1159	proportional-plus-derivative action	action f proportionnelle et par dérivation (PD)	PD-Verhalten	пропорционально-дифференцирующее воздействие	azione proporzionale-derivativa	acción f proporcional y derivada	比例＋微分動作 hirei+bibun-dōsa	比例微分作用 bili weifen zuoyong
1160	proportional-plus-integral action	action f proportionnelle et par intégration (P I)	PID-Verhalten	ПИД-регулятор	azione proporzionale-integrale	acción f proporcional e integral, acción f proporcional e integral	比例＋積分動作 hirei+sekibun-dōsa	比例积分作用 bili jifen zuoyong
1161	proportional-plus-integral controller	régulateur m proportionnel et intégral	PI-Verhalten	ПИД-воздействие	controllore proporzionale-integrale	regulador m de acción proporcional e integral	比例＋積分＋調整器 hirei+sekibun-chōsetsuki	比例积分控制器 bili jifen kongzhiqi
1162	proportional-plus-integral-derivative action	action f PID	PID-Regler, Regler mit proportionalem,integrierendem u. differenzierendem halten	пропорционально-интегрирующее воздействие	azione proporzionale-integrale-derivativa	acción f proporcional, integral y derivada, acción f proporcional, integral y derivada	比例＋積分＋微分動作 hirei+sekibun+bibun-dōsa	比例积分微分作用 bili jifen weifen zuoyong
1163	proportional-plus-integral-derivative controller	régulateur m proportionnel par intégration et par dérivation	PI-Regler, Regler mit proportionalem u. integrierendem Verhalten	ПИ-регулятор, изодромный регулятор	controllore proporzionale-integrale-derivativo	regulador m de acción proporcional, integral y derivada	比例＋積分＋微分調整器 hirei+sekibun+bibun-chōsetsuki	比例积分微分控制器 bili jifen weifen konzhiqi
1164	pseudo-random sequence	suite f, séquence f binaire pseudo-aléatoire	Pseudozufallsfolge	псевдослучайная последовательность	sequenza pseudocasuale	secuencia * seudoaleatoria	疑似ランダム数列 giji-randmu-suretsu	伪随机序列 weisuiji xulie
1165	pulse	impulsion f	Impuls	импульс	impulso	impulso m, pulso m	パルス parusu	脉冲 maichong
1166	pulse (train) function	train m d'impulsions	Impulsfolgefunktion	решётчатая функция	funzione treno d'impulsi	función f de un tren de impulsos	パルス（列）関数 parusu(retsu)-kansū	脉冲（序列）函数 maichong (xulie) hanshu
1167	pulse signal	signal m impulsionnel	Impulssignal	импульсный сигнал	segnale impulso	señal f de impulsos	パルス信号 parusu-shingō	脉冲信号 maichong xinhao
1168	pulse train	train m d'impulsions	Impulsfolge	последовательность импульсов, серия импульсов	treno d'impulsi	tren m de impulsos	パルス列 parusu-retsu	脉冲序列 maichong xulie
1169	pulse width	largeur f d'impulsion	Pulsbreite, Impulsdauer	длительность импульса, ширина импульса	larghezza dell'impulso	anchura f de impulso, duración f de impulso	パルス幅 parusu-haba	脉冲宽度 maichong kuandu
1170	pump	pompe f	Pumpe	насос	pompa	bomba f	ポンプ ponpu	泵 beng

№	English	French	German	Russian	Italian	Spanish	Japanese	Chinese
1171	push-pull amplifier	amplificateur *m* symétrique	Gegentaktverstärker	двухтактный усилитель	amplificatore "push-pull"	amplificador *m* en contrafase, amplificador *m* "push-pull"	プッシュプル増幅器 pusshupuru-zōfukuki	推挽放大器 tuīwǎn fàngdàqì
1172	quadratic performance index	indice *m* de qualité quadratique	quadratisches Gütekriterium, quadratischer Güteindex	квадратичный критерий качества	cifra di merito quadratica	índice *m* cuadrático de funcionamiento	自乗評価指数 jijyō-hyōka-shisū	二次性能指标 èrcìxìng xìngnéng zhǐbiāo
1173	quadrature axis	axe *m* en quadrature	Querachse, Querfeldachse	перпендикулярные оси, оси под прямым углом	asse in quadratura	eje en cuadratura	横軸 yoko-jiku	正交轴 zhèngjiāozhóu
1174	quadrature brushes	balais *m* en quadrature	Querbürsten	щетки под прямым углом	spazzole in quadratura	escobillas de cuadratura	横軸ブラシ yoko-jiku-burashi	正交电刷 zhèngjiāo diànshuā
1175	quadripole	quadripôle *m*	Vierpol	четырехполюсник	quadripolo	cuadripolo *m*	4次極 yoji-kyoku	四端网络 sìduān wǎngluò
1176	quality control	gestion *f* de la qualité	Qualitätskontrolle	управление качеством	controllo di qualità	control *m* de calidad	品質管理 hinshitsu-kanri	质量控制 zhìliàng kòngzhì
1177	quantity	variable *f*	Menge, Größe (math.), Quantum	количество	quantità	cantidad	量	量，数量 liàng, shùliàng
1178	quantization	quantification *f*	Quantisierung	квантование	quantizzazione	cuantificación *f*	量子化 ryōshika	量化 liànghuà
1179	quantization error	erreur *f* de quantification	Quantisierungsfehler	ошибка квантования	errore di quantizzazione	error *m* de cuantificación	量子化誤差 ryōshika-gosa	量化误差 liànghuà wùchā
1180	quantized signal	signal *m* quantifié	quantisiertes Signal	квантованный (по уровню) сигнал	segnale quantizzato	señal *f* cuantificada	量子化信号 ryōshika-shingō	量化信号 liànghuà xìnhào
1181	queuing theory	théorie *f* des files d'attente	Bedienungstheorie, Warteschlangentheorie	теория очередей	teoria delle code	teoría *f* de colas	待合せ理論 machi-gyōretsu-riron	排队论 páiduìlùn
1182	rack-type differential	différentiateur *m* normalisé insérable	Zahnstangendifferential	зубчатый дифференциал	differenziale a cremagliera	diferencial tipo-rack	ラック形差動機構 rakkugata-sadō-haguruma	齿条差动机构 chǐtiáo chādòng jīgòu
1183	radial network (in decentralized control)	réseau *m* radial (en régulation répartie)	strahlungsförmiges Netzwerk (bei dezentraler Regelung)	радиальная сеть	rete radiale (nel controllo decentralizzato)	red *f* en estrella, red *f* radial	星形線網路 hoshigata-senrōmō	(分散控制的)径向网络 (fēnsànkòngzhìde) jìngxiàng wǎngluò

No.	English	French	German	Russian	Italian	Spanish	Japanese	Chinese
1184	radial pump	pompe f radiale	Radialpumpe	радиальный насос	pompa radiale	bomba f radial	ラジアルポンプ rajiaru-ponpu	径向泵 jingxiangbeng
1185	ram	vérin m	Ramme, Preßkolben, Widder	плунжер	rampa	vástago m, pistón m	ラム ramu	作动筒、柱塞 zuodongtong, zhusai
1186	ramp function	fonction f rampe, échelon m de vitesse	Rampenfunktion	ступенчатая функция	funzione rampa	(función) rampa	ランプ関数 ranpu-kansū	斜坡函数 xiepo hanshu
1187	ramp function response	réponse f à la fonction rampe	Rampenfunktionsantwort	реакция на линейно возрастающий вход	risposta a rampa	respuesta f a una rampa	ランプ関数応答 ranpu-kansū-ōtō	斜坡函数响应 xiepo hanshu xiangying
1188	ramp input	grandeur f d'entrée en forme de rampe, échelon m de vitesse	rampenförmige Eingangsgröße	линейно возрастающий вход	ingresso a rampa	entrada f en rampa	ランプ入力 ranpu-nyūryoku	斜坡输入 xiepo shuru
1189	random	aléatoire	ufällig, regellos, wahllos, statistisch verteilt	случайный	casuale	aleatorio	不規則な fukisoku-na	随机的 suijide
1190	random access memory (RAM)	mémoire f vive, à accès aléatoire	Direktzugriffsspeicher, Speicher mit wahlfreiem Zugriff (RAM)	память с произвольным доступом, оперативная память	memoria ad accesso casuale (RAM)	memoria * de acceso aleatorio	高速呼び出し記憶装置 tōsoku-yobidashi-kioku-sōchi	随机存取储器 suijicunqu cunchuqi
1191	random drift	dérive f aléatoire	Geschwindigkeit der Zufallsauswanderung	случайный дрейф	deriva erratica	deriva f aleatoria	不規則ドリフト fukisoku-dorifuto	随机漂移 suiji piaoyi
1192	random function	fonction f aléatoire	Zufallsfunktion	случайная функция	funzione casuale	Booleana * aleatoria	不規則関数 randamu-kansū	随机函数 suiji hanshu
1193	random number generator	générateur m de nombres aléatoires	Zufalls(zahlen)generator	генератор случайных чисел	generatore di numeri casuali	generador * de números aleatorios	乱数発生器 ransū-hasseiki	随机数发生器 suijishu fashengqi
1194	random process	fonction f aléatoire	Zufallsprozeß, stochastischer Prozeß	случайный процесс	processo casuale	proceso m aleatorio	不規則過程 randamu-purosesu	随机过程 suiji guocheng
1195	random variable	variable f aléatoire	Zufallsvariable, Zufallsgröße	случайная переменная	variabile casuale	variable f aleatoria	不規則変数 fukisoku-hensū	随机变量 suiji bianliang
1196	range	domaine m, étendue f	Bereich, Aufzeichnungsbereich, Wertevorrat (einer Funktion)	диапазон	gamma	margen m	範囲 han-i	范围、量程 fanwei, liangcheng

No.	English	French	German	Russian	Italian	Spanish	Japanese	Chinese
1197	range of disturbance	domaine *m* de perturbation	Störbereich	диапазон изменения возмущений (помех)	gamma dei disturbi	margen *m* de perturbación	外乱範囲 gairan-han-i	扰动范围 raodong fanwei
1198	range of set value	domaine *m* de consigne	Arbeits(punkt)bereich	область заданных значений	gamma dei valori d'insieme	gama *f* de valores de consigna	設定範囲 settei-han-i	设定值范围 shedingzhi fanwei
1199	range splitting	division *f* de gamme	Bereichsaufteilung	расщепление диапазона	partizione di gamma	división *f* de escala		范围划分 fanwei huafen
1200	rank	rang *m*	Rang	ранг	rango	rango *m*	階 級 ランク kaikyū ranku	秩、排、级 zhi, pai, ji
1201	rate action	action *f* par dérivation, action *f* D	Differentialverhalten, Vorverhalten	воздействие по скорости, воздействие по производной	azione sulla velocità	acción *f* derivada, acción *f* D	レート動作 rēto-dōsa	速率作用 sulü zuoyong
1202	rate feedback	réaction *f* tachymétrique	differenzierende Rückführung	обратная связь по скорости	retroazione di velocità	realimentación *f* taquimétrica	微分フィードバック bibun-fido-bakku	速率反馈 sulü fankui
1203	rate gyro	gyromètre *m*	Meßkreisel, Wendekreisel	скоростной гироскоп	giroscopio di velocità	girómetro *m*	レートジャイロ rēto-jairo	速率陀螺 sulü tuoluo
1204	rate of inherent regulation	vitesse *f* d'auto-régulation	Rate der Spannungs- oder Drehzahlän-derung (siehe. inherent regulation)	скорость саморегулирования	tasso di regolazione intrinseca	velocidad de regulación inherente		固有调节速率 guyou tiaojie sulü
1205	rate of self-regulation	vitesse *f* d'auto-régulation	Rate der Selbstregulierung od. der selbständigen Regelung	скорость саморегулирования	tasso di autoregolazione	velocidad de auto regulación	自己平衡性割合 jiko-heikōsei-wariai	自调节速率 zitiaojie sulü
1206	ratio control	commande *f* de proportion	Verhältnisregelung	регулирование соотношения	controllo di rapporto	control *m* de relación	比率制御 hiritsu-seigyo	比值控制 bizhi kongzhi
1207	reachability	accessibilité *f*	Erreichbarkeit	достижимость	raggiungibilità	alcanzabilidad	可到達性 ka-tōtatsu-sei	可达性 kedaxing
1208	reachable state	état *m* accessible	erreichbarer Zustand	достижимое состояние	stato raggiungibile	estado *m* alcanzable	可到達状態 ka-tōtatsu-jōtai	可达状态, 能达状态 kedazhuangtai, nengdazhuangtai
1209	real axis	axe *m* réel	reele Achse	действительная ось	asse reale	eje *m* imaginario	実軸 jitsu-jiku	实轴 shizhou

No.	English	French	German	Russian	Italian	Spanish	Japanese	Chinese
1210	real-time	temps-réel *m*	Echtzeit	реальное время, в реальном масштабе времени	tempo-reale	tiempo real	実時間 jitsu-jikan	实时 shishi
1211	real-time computer	calculateur *m* en temps-réel	Echtzeitrechner	вычислительная машина, работающая в реальном масштабе времени	calcolatore in tempo reale	computador *m* de tiempo real	実時間計算機 jitsujikan-keisanki	实时计算机 shishi jisuanji
1212	real-time language	langage *m* en temps-réel	Echtzeitsprache	язык реального времени	linguaggio (per applicazioni) in tempo reale	lenguaje *m* de tiempo real	実時間言語 jitsujikan-gengo	实时语言 shishi yuyan
1213	real-time system	système *m* en temps-réel	Echtzeitsystem	система реального времени	sistema in tempo reale	sistema *m* de tiempo real	実時間系（システム） jitsujikan-kei (-shisutemu)	实时系统 shishi xitong
1214	reconstructability	reconstructibilité *f*	Rekonstruktierbarkeit	перестраиваемость	ricostruibilità	reconstructibilidad *f*	可決定性 ka-kettei-sei	可重建性 kechongjianxing
1215	recovery time	temps *m* de retour à la normale, de recouvrement	Erholzeit, Ausgleichszeit	время восстановления	tempo di recupero, o di riattivazione	tiempo *m* de recuperación	回復時間 kaifuku-jikan	恢复时间，过渡过程时间 huifu shijian, guodu guocheng shijian
1216	rectangular wave	onde *f* rectangulaire	Rechteckschwingung	прямоугольный сигнал	onda rettangolare	onda *f* rectangular	方形波 hōkei-ha	矩形波 juxingbo
1217	recursive control algorithm	algorithme *m* de réglage récurrent	rekursiver Regelungs- od. Steuerungsalgorithmus	рекурсивный алгоритм управления	algoritmo di controllo ricorsivo	algoritmo *m* recursivo de control	再帰的制御アルゴリズム saikiteki-seigyo-arugorizumu	递归(推)控制算法 digui kongzhi suanfa
1218	reduced-order model	modèle *m* d'ordre réduit	Modell reduzierter Ordnung	модель пониженного порядка	modello d'ordine ridotto	modelo *m* de orden reducido	低次元モデル teijigen-moderu	降阶模型 jiangjie moxing
1219	reducing valve	détendeur *m*	Reduzierventil, Druckminderungsventil	редукционный клапан	valvola di riduzione	válvula *f* reductora	減圧弁 gen-atsu-ben	减压阀 jianyafa
1220	redundancy	redondance *f*	Redundanz	резервирование, избыточность	ridondanza *f*	redundancia *f*	冗長度 jōchōdo	冗余，复式 rongyu, fushi
1221	reference element	organe *m* de référence	Vergleichselement	опорный элемент	generatore del riferimento	elemento *m* de referencia	目標値要素 mokuhyōchi-yōso	参考环节，参考元件 cankao huanjie, cankao yuanjian
1222	reference-input element	générateur *m* de grandeur de référence	Führungssignalgeber	опорный элемент	generatore dell'ingresso di riferimento	elemento *m* de referencia	基準入力要素 kijyun-nyūryoku-yōso	参考输入环节 cankao shuru huanjie

No.	English	French	German	Russian	Italian	Spanish	Japanese	Chinese
1223	reference-input signal	signal *m* de référence	Führungssignal	задающее воздействие, задающий сигнал, опорный сигнал	segnale ingresso di riferimento	señal *f* de referencia, señal *f* patron	基準入力信号 kijyun-nyūryoku-shingō	参考输入信号 cankao shuru xinhao
1224	reference-input variable	grandeur *f* d'entrée de référence	Führungsgröße, Leitsollwert	задающее воздействие	variabile ingresso di riferimento	magnitud *f* piloto, variable *f* de referencia	基準入力変数 kijyun-nyūryoku-hensū	参考输入变量 cankao shuru bianliang
1225	reference signal	signal *m* de référence	Bezugssignal	задающее воздействие, задающий сигнал, опорный сигнал	segnale di riferimento	señal *f* de referencia, señal *f* patron	目標値信号 mokuhyo-chi-shingō	参考信号 cankao xinhao
1226	reference variable	grandeur *f* de référence	Bezugsgröße	задающее воздействие	variabile di riferimento	magnitud *f* piloto, variable *f* de referencia		参考变量 cankao bianliang
1227	reference winding	enroulement *m* de référence	Referenzwicklung	задающая обмотка	avvolgimento di riferimento	arrollamiento *m* de referencia, devanado *m* de referencia	基準巻線 kijun-makisen	基准绕组, 参考绕组 jizhun raozu, cankao raozu
1228	regeneration	remise *f* en forme, régénération *f*	Regeneration, Wiedergewinnung	регенерация	rigenerazione	regeneración *	再生 saisei	再生 zaisheng
1229	regenerative feedback	réaction *f* positive	Mitkopplung, positive Rückkopplung	положительная обратная связь	retroazione rigenerativa	realimentación *f* positiva		正反馈 zhengfankui
1230	register	registre *m*	Register, Zähleinrichtung	регистр	registro	registro *m*	計数器 レジスタ chisūki rejisuta	寄存器 jicunqi
1231	regulate (v)	régler (v)	regulieren, einstellen, regeln	регулировать	regolare	regular	制御する seigyo-suru	调节 (动词) tiaojie (dongci)
1232	regulating element	organe *m* de régulation	Reguliervorrichtung, Stellglied	исполнительный орган	elemento di regolazione	elemento *m* de regulación	調節要素 chōsetsu-yōso	调节环节, 调节元件 tiaojie huanjie, tiaojie yuanjian
1233	regulating energy	énergie *f* réglante	Stellenergie	энергия регулирования	energia di regolazione	energía *f* de regulación	調整エネルギー chōsei-enerugi	调节能量 tiaojie nengliang
1234	regulating unit	organe *m* de régulation	Reguliereinrichtung, Stellorgan	исполнительный орган	unità di regolazione	unidad *f* de regulación	調節要素 chōsetsu-yōso	调节单元 tiaojie danyuan
1235	regulation	régulation *f*	Regulierung, Regelung, Regel, Vorschrift	регулирование; статизм	regolazione	regulación *f*	変動率 hendō-ritsu	调节 tiaojie

No.	English	Français	Deutsch	Русский	Italiano	Español	日本語	中文
1236	regulator	régulateur m	Regulator, Regler, Konstanthalter	регулятор	regolatore	regulador m	調整器 chōseiki	调节器 tiaojieqi
1237	relative control range	étendue f relative de réglage	relativer Regelbereich	относительный диапазон регулирования	gamma di controllo relativa	margen m relativo de control	相対的制御範囲 sōtaiteki-seigyo-han-i	相对控制范围 xiangdui kongzhi fanwei
1238	relaxation	relaxation f	Relaxation, Entspannung, Kippschwingung	релаксация	rilassamento	relajación f	緩和 kanwa	松弛 songchi
1239	relay	relais m	Relais	реле	relay	relé m	継電器 リレー keidenki rirē	继电器 jidianqi
1240	release (v)	déverrouiller (v)	auslösen, freigeben	освобождать	rilasciare	desaccionar, liberar	開放する kaihō-suru	释放, 脱扣 (动词) shifang, tuokou (dongci)
1241	reliability	fiabilité f	Zuverlässigkeit	надежность	affidabilità	fiabilidad f	信頼性 shinraisei	可靠性, 可靠度 kekaoxing, kekaodu
1242	reliable	fiable	zuverlässig	надежный	affidabile	fiable, seguro	信頼的 shinrai-dekiru	可靠的 kekaode
1243	relief valve	soupape f de sûreté	Sicherheitsventil, Überdruckventil, Entlastungsventil	предохранительный клапан, разгрузочный клапан	valvola di rilascio	válvula f de descarga, válvula f de seguridad	リリーフ弁 rirīfu-ben	安全阀 anquanfa
1244	remote control (open loop)	télécommande f	Fernsteuerung, Fernbedienung	телеуправление (разомкнутое)	controllo a distanza (ad anello aperto)	control m remoto (anillo abierto), telemando m	遠隔制御 enkaku-seigyo	遥控 (开环) yaokong (kaihuan)
1245	repeater motor	moteur m pas-à-pas	Wiederholermotor	шаговый двигатель	motore ripetitore	motor m paso a paso		步进电机 bujin dianji
1246	reproducibility	reproductibilité f, fidélité f	Reproduzierbarkeit	воспроизводимость	riproducibilità	reproducibilidad f	再現性 saigensei	再现性 zaixianxing
1247	reproducible	reproductible	reproduzierbar, nachbildbar	воспроизводимый	riproducibile	reproducible	再現できる saigen-dekiru	可再现的 kezaixiande
1248	reset action	action f intégrale	Integralverhalten	пропорционально-интегрирующее воздействие	azione di riassetto (azione integrale)	acción f proporcional e integral, acción f proporcional e integral	リセット動作 risetto-dōsa	重复作用, 重调作用 chongzhi zuoyong, chongtiao zuoyong

No.	English	French	German	Russian	Italian	Spanish	Japanese	Chinese
1249	reset time	temps *m* (de dosage) d'intégration	Nachstellzeit, Integralzeit, Rückstellzeit	время возврата (в исходное положение)	tempo di riassetto (tempo integrale)	tiempo *m* de restitución, tiempo *m* de reposición	リセット時間 risetto-jikan	重置时间, 复调时间 chongzhi shijian, chongtiao shijian
1250	resistor	résistance *f*	Widerstand	резистор, сопротивление	resistore	resistencia *f*, resistor *m*	抵抗器 teikōki	电阻器 dianzuqi
1251	resolution	résolution *f*	Auflösung (svermögen), Feinheit (der Ein- u. Ausgabe)	разрешающая способность	risoluzione	resolución *f*	分解能 bunkainō	分辨率 fenbianlü
1252	resolvent matrix	matrice *f* résolvante	Resolventenmatrix	матрица резольвенты	matrice risolvente	matriz *f* resolvente		預解矩陣 yujie juzhen
1253	resonance	résonance *f*	Resonanz, Nachhall	резонанс	risonanza	resonancia *f*	共振 kyōshin	谐振 xiezhen
1254	resonance ratio	facteur *m* de résonance	Resonanzüberhöhung	относительный резонансный пик	rapporto di risonanza	coeficiente *m* de resonancia	共振比 kyōshin-hi	谐振比 xiezhenbi
1255	resonant frequency	fréquence *f* de résonance	Resonanzfrequenz	резонансная частота	frequenza (o pulsazione) di risonanza	frecuencia *f* de resonancia	共振周波数 kyōshin-shūhasū	谐振频率 xiezhen pinlü
1256	response	réponse *f*	Antwort, Reaktion, Verhalten, Dynamik	реакция	risposta	respuesta *f*	応答 ōtō	响应 xiangying
1257	response curve	courbe *f* de l'écart de réglage	(Ansprech)kennlinie, Frequenzkurve, Wiedergabekurve	кривая реакция	curva di risposta	curva *f* de respuesta	応答曲線 ōtō-kyokusen	响应曲线 xiangxing quxian
1258	response/frequency diagram	diagramme *m* de Bode	Frequenzgang (Ausgangsgröße)	диаграмма Боде	diagramma della risposta in frequenza	diagrama *m* de respuesta de frecuencia	応答／周波数線図 ōtō/shūhasū-senzu	频率响应图 pinlü xiangyingtu
1259	response function	réponse *f* temporelle	Antwortfunktion, Ausgangsfunktion	переходная функция	(funzione di) risposta	función *f* de respuesta	応答関数 ōtō-kansū	响应函数 xiangying hanshu
1260	response time	temps *m* de réponse	Einstellzeit, Einstelldauer	время установления	tempo di risposta	tiempo *m* de respuesta	応答時間 ōtō-jikan	响应时间 xiangying shijian
1261	rest duration	durée *f* de repos	Stillstandsdauer	длительность паузы	durata del (periodo di) riposo	duración *f* de reposo	休息時間 kyūsoku-jikan	休止时间 xiuzhi shijian

	English	French	German	Russian	Italian	Spanish	Japanese	Chinese
1262	restart	redémarrage m	Wiederanlauf, Wiedereinschaltung	рестарт, повторный запуск	riavvio	rearranque m, reinicio m	リスタート 再起動 risutāto saikidō	再开动 zaikaidong
1263	restrictor	étranglement m	Drossel(stelle), (Meß)blende, Durchflußbegrenzer	ограничитель	restrittore	limitador m, reductor m	絞り sibori	节减器, 限制器 jieliufa, xianzhiqi
1264	return difference determinant	déterminant m de la différence de retour	Rückführdifferenz-Determinante	определитель возвратных разностей	determinante della differenza di ritorno	determinante m de la diferencia de retorno	還送差行列式 kansōsa-gyōretsu	回差行列式 huicha hanglieshi
1265	return difference ratio	rapport m de retour du signal de différence	Rückführdifferenz-Verhältnis	отношение возвратных разностей	rapporto della differenza di ritorno	relación de la diferencia realimentada	還送差比 kansōsa-hi	回差比 huichabi
1266	return signal	signal m de réaction	Rückführsignal, Rückmeldung, Echosignal	возвратный сигнал	segnale di ritorno	señal f de realimentación, señal f de retorno	返回信号 fanhui xinhao	返回信号 fanhui xinhao
1267	return transfer function	transmittance f de la chaîne de réaction	Übertragungsfunktion des Rückführpfades	обратная передаточная функция	funzione di trasferimento di ritorno	función f de transferencia del canal de realimentación, función f de transferencia de la cadena de retorno	帰還伝達関数 kikan-dentatsu-kansū	返回传递函数 fanhui chuandi hanshu
1268	reversible system	système m réversible	umkehrbares (reversibles) System	реверсивная система	sistema reversibile	sistema f reversible	可逆系 kagyaku-kei	可逆系统 keni xitong
1269	ring network (in decentralized control)	réseau m annulaire (en régulation répartie)	Ringnetz, ringförmiges Netz (dezentrale Regelung od. Steuerung)	кольцевая сеть (при децентрализованном управлении)	rete ad anello (nel controllo decentralizzato)	red f en anillo	環状網路線 kanjō-senromō	(分散控制的)环形网络 (fensankongzhide) huanxing wangluo
1270	rise time	temps m de montée	Anstiegszeit, Anregelzeit	время нарастания	tempo di salita	tiempo m de subida	立ち上がり時間 tachiagari-jikan	上升时间 shangsheng shijian
1271	robot	robot m	Roboter, automatische Vorrichtung	робот	robot	robot m	ロボット robotto	机器人 jiqiren
1272	robotics	robotique f	Robotertechnik, Handhabungstechnik, Robotik	робототехника	robotica	robótica f	ロボティクス robotikusu	机器人学 jiqirenxue
1273	robust control	régulation f robuste	robuste Regelung	робастное управление, грубое управление	controllo robusto	control m robusto	ロバスト制御 robasuto-seigyo	鲁棒(稳健)控制 lubang kongzhi
1274	robust estimator	estimateur m robuste	robuster Schätzer	робастный оцениватель	stimatore robusto	estimador m robusto	ロバスト推定器 robasuto-suiteiki	鲁棒(稳健)估计器 lubang gujiqi

No.	English	French	German	Russian	Italian	Spanish	Japanese	Chinese
1275	robustness	robustesse f	Robustheit	робастность, грубость	robustezza	robustez f	頑健さ ロバストネス ganken-sa robasutonesu	鲁棒性, 稳健性 lubangxing
1276	root	racine f	Wurzel (Lösung eines Polynoms), Wurzel, Fußpunkt	корень	radice	raíz f	根 kon	根 gen
1277	root locus	lieu m des pôles, lieu m des racines	Wurzelort(skurve)	корневой годограф	luogo delle radici	lugar m (geométrico) de las raíces	根軌跡 kon-kiseki	根轨迹 genguiji
1278	root locus diagram	lieu m des racines	Wurzelortskurve (ndarstellung)	диаграмма корневого годографа	diagramma del luogo delle radici	diagrama del lugar de las raíces	根軌跡線図 kon-kiseki-senzu	根轨迹图 genguijitu
1279	root-mean-square value	valeur f moyenne quadratique	quadratischer Mittelwert, Effektivwert	среднеквадратичное значение	radice quadrata del valore quadratico medio	valor m eficaz, valor m medio cuadrático	2乗根平均値 jijyōkon-heikin-chi	均方根值 junfanggenzhi
1280	rotor	rotor m	Rotor	ротор	rotore	rotor m	回転子 kaitenshi	转子 zhuanzi
1281	Routh's criterion	critère m de Routh	Routh-Kriterium	критерий устойчивости Рауса	criterio di Routh	criterio m de Routh	ラウスの(安定)基準 Routh-no-(antei)kijun	劳斯判据 laosi panju
1282	run time system	temps m d'exécution d'un système	Laufzeitsystem	система (измерения) времени прогона (программы)	sistema 'run time'	sistema de ejecución		运行时间系统 yunxing shijian xitong
1283	Runge-Kutta method	méthode f de Runge-Kutta	Runge-Kutta-Verfahren	метод Рунге-Кутта	metodo di Runge-Kutta	método de runge-kutta	ルンゲクッタ法 Runge-Kutta-hō	龙格-库塔法 longge-kutafa
1284	saddle	col m	Sattel	седловина	sella	puerto m	鞍状点 anjōten	鞍, 鞍点 an, andian
1285	sample	échantillon m	Abtastung, (Stich-)Probe, Muster	единичный замер, выборка	campione	muestra f	サンプル sanpuru	采样, 样本 caiyang, yangben
1286	sample (v)	échantillonner (v)	abtasten, Stichprobe entnehmen	квантовать во времени	campionare	muestrear m	サンプルする sanpuru-suru	采样 (动词) caiyang (dongci)
1287	sample-and-hold	échantillonné-bloqué	Halteglied (Abtastregelung)	аналоговое запоминание	campionamento e tenuta (organo di)	muestreo m y retención	サンプルホールド sanpuru-hōrudo	采样保持 caiyangbaochi

No.	English	French	German	Russian	Italian	Spanish	Japanese	Chinese
1288	sampled data	données f,pl échantillonnées	Abtastwerte	квантованная величина	segnali (dati) campionati	datos m muestreados	サンプル値 sanpuruchi	采样数据 caiyang shuju
1289	sampled-data control	commande f à échantillonnage	Abtastregelung	дискретное управление	controllo a segnali campionati	control m muestreado	サンプル値制御 sanpuruchi-seigyo	采样控制 caiyang kongzhi
1290	sampled data system	système m échantillonné	Abtastsystem	дискретная система	sistema a segnali campionati	sistema * muestreado	サンプル値系 sanpuruchi-kei	采样系统 caiyang xitong
1291	sampled signal	signal m échantillonné	abgetastetes Signal	дискретный (по времени) сигнал	segnale campionato	señal f muestreada	サンプル値信号 sanpuruchi-singō	采样信号 caiyang xinhao
1292	sampler	échantillonneur m	Abtaster	элемент выборки, импульсный элемент	campionatore	muestreador m	サンプラ sanpura	采样器 caiyangqi
1293	sampling	échantillonnage m	Abtastung	квантование во времени	campionamento	muestreo m	サンプリング sanpuringu	采样 caiyang
1294	sampling action	action f d'échantillonnage	Abtastvorgang	периодическое действие, дискретизирующее действие, импульсное действие	azione di campionamento	acción f de muestro	サンプリング動作 sanpuringu-dōsa	采样作用 caiyang zuoyong
1295	sampling control	commande f à échantillonnage	Abtastverhalten (Regelung od. Steuerung)	дискретное управление	controllo a campionamento	control m muestreado	サンプル値制御 sanpuruchi-seigyo	采样控制 caiyang kongzhi
1296	sampling controller	régulateur m à action intermittente	Abtastregler	импульсный регулятор	controllore a campionamento	controlador m de muestreo	サンプリング調節器 sanpuringu-chōsetsuki	采样控制器 caiyang kongzhiqi
1297	sampling element	échantillonneur m	Abtastelement	элемент выборки, импульсный элемент	elemento campionatore	elemento m de muestreo	サンプリング要素 sanpuringu-yōso	采样元件 caiyang yuanjian
1298	sampling frequency	fréquence f d'échantillonnage	Abtastfrequenz	частота квантования	frequenza (pulsazione) di campionamento	frecuencia f de muestreo	サンプリング周波数 sanpuringu-shūhasū	采样率 caiyang pinlü
1299	sampling interval	intervalle m d'échantillonnage	Abtastinterval	интервал выборки	intervallo di campionamento	intervalo m de muestreo	サンプリング間隔 sanpuringu-kikan	采样间隔 caiyang jiange
1300	sampling period	période f d'échantillonnage	Abtastperiode	период квантования	periodo di campionamento	periodo m de muestreo	サンプリング周期 sanpuringu-shūki	采样周期 caiyang zhouqi

	English	Français	Deutsch	Русский	Italiano	Español	日本語	中文
1301	sampling rate	cadence f d'échantillonnage	Abtastrate, Abtastgeschwindigkeit	частота квантования	velocità di campionamento	cadencia f de muestreo	サンプリング周波数	采样率 caiyanglü
1302	sampling system	système m d'échantillonnage	Abtastsystem	импульсная система	sistema di campionamento	sistema m de muestreo	サンプリングシステム sanpuringu-shisutemu	采样系统 caiyang xitong
1303	saturation	saturation f	Sättigung	насыщение	saturazione	saturación f	飽和 hōwa	饱和 baohe
1304	scanning control	commande f à balayage	Abtaststeuerung (Rechengerät), Bildabtaststeuerung	сканирующее управление	controllo a scansione	control m por barrido	走査制御 sōsa-seigyo	扫描控制 saomiao kongzhi
1305	schematic diagram	schéma m des liaisons	schematisches Diagramm, schematische Darstellung	структурная схема	diagramma schematico	diagrama m esquemático	構成図	原理图 yuanlitu
1306	searching system	système m de recherche	Suchsystem	поисковая система	sistema di puntamento automatico	sistema m de exploración	探索システム tansaku-shisutemu	搜索系统 sousuo xitong
1307	second-derivative action	action f par double dérivation	Auswirkung der zweiten Ableitung	воздействие по второй производной	azione alla derivata seconda	acción f derivada doble	二次微分動作 niji-bibun-dōsa	二阶微分作用 erjie weifen zuoyong
1308	second-order system	système m du deuxième ordre	System zweiter Ordnung	система второго порядка	sistema del second'ordine	sistema m de segundo orden	二次系 niji-kei	二阶系统 erjie xitong
1309	secondary regulation	régulation f secondaire	Sekundärregulation, unterlagerte Steuerung	вторичное регулирование	regolazione secondaria	regulación f secundaria	二次調節 niji-chōsetsu	二次调节 erci tiaojie
1310	selector switch	sélecteur m, commutateur m	Wahlschalter, Meßstellenumschalter	селекторный переключатель	commutatore di selezione	conmutador m, conmutador m selector	切り換えスイッチ kirikae-suicchi	选择开关 xuanze kaiguan
1311	self-acting control (closed loop)	régulation f directe	selbsttätige Regelung	прямое (непосредственное) управление (регулирование)	controllo auto-attivante (ad anello chiuso)	regulación f autónoma	自力制御	自作用控制(闭环) zizuoyong kongzhi (bihuan)
1312	self-adapting algorithm	algorithme m auto-adaptatif	selbstadaptierender Algorithmus	самонастраивающийся алгоритм	algoritmo auto-adattativo	algoritmo m autoadaptativo	自己適応アルゴリズム jiko-teikō-arugorizumu	自适应算法 zishiying suanfa
1313	self-adaptive control	commande f auto-adaptative	selbsttätig abgleichende Regelung, selbstadaptierende Regelung	самонастраивающееся (адаптивное) управление	controllo auto-adattativo	control * autoadaptativo	自己適応制御 jiko-teikō-seigyo	自适应控制 zishiying kongzhi

	English	French	German	Russian	Italian	Spanish	Japanese	Chinese
1314	self-adjusting system	système m auto-adaptatif	selbsteinstellendes System	самонастраивающаяся система	sistema a taratura automatica	sistema m autoajustable	自己調整系 (システム) jiko-chōsei-kei (-shisutemu)	自调整系统 zitiaozheng xitong
1315	self-aligning control (closed loop)	régulation f de correspondance	selbstabgleichende Regelung	самоустанавливающееся управление, управление с жёсткой обратной связью, следящее регулирование	controllo ad auto-allineamento, (ad anello chiuso)	control m de seguimiento (en anillo cerrado), servocontrol m		自对准控制 (闭环) ziduizhun kongzhi (bihuan)
1316	self-excitation winding	enroulement m d'auto-excitation	Selbsterregerwicklung	обмотка самовозбуждения	avvolgimento di autoeccitazione	devanado m de autoexcitación	自己励磁巻線 jiko-reishin-makisen	自激绕组 ziji raozu
1317	self-excited oscillation	auto-oscillation f	selbstanregende Oszillation od. Schwingung	автоколебания	oscillazione autoinnescata	autooscilación f, oscilación f autoentretenida	自励振動 jirei-shindō	自激振荡 ziji zhendang
1318	self-operated control	régulation f directe	Regelung ohne Hilfsenergie	прямое (непосредственное) управление (регулирование)	controllo ad azionamento automatico	regulación f autónoma	自力制御 jiriki-seigyo	自力式控制 zilishi kongzhi
1319	self-optimizing control	régulation f auto-optimisante	selbstoptimierende Regelung	экстремальное управление (регулирование)	controllo auto-ottimizzante	control m autooptimizante	自己最適化制御 jiko-saitekika-seigyo	自寻优控制 zixunyou kongzhi
1320	self-optimizing system	système m auto-adaptatif, auto-optimisant	selbstoptimierendes System	система экстремального управления	sistema auto-ottimizzante	sistema auto-optimizante	自己最適化システム jiko-saitekika-shisutemu	自寻优系统 zixunyou xitong
1321	self-organizing storage	stockage m auto-organisé	selbstorganisierende Speicherung	самоорганизующаяся память	memoria auto-organizzantesi	almacenamiento m autoorganizativo		自组织存储器 zizuzhi cunchuqi
1322	self-organizing system	système m auto-organisé	selbstorganisierendes System	самоорганизующаяся система	sistema auto-organizzantesi	sistema m autoorganizativo	自己組織系 jiko-keisei-kei (-shisutemu)	自组织系统 zizuzhi xitong
1323	self-oscillation	auto-oscillation f	Eigenschwingung	автоколебания	auto-oscillazione	autooscilación f, oscilación f autoentretenida	自励振動 jirei-shindō	自振荡 zizhendang
1324	self-regulation	auto-régulation f	Selbstregulierung, automatische od. selbsttätige Steuerung	саморегулирование	auto-regolazione	autorregulación f	自己平衡性 jiko-heikōsei	自调整 zitiaozheng
1325	self-reproducing automata	automates m pl auto-reproducteurs	selbstreproduzierende Automaten	самовоспроизводящийся автомат	automa autoriproducentesi	autómata m autoreproductor	自己再生オートマタ jiko-saisei-ōtomata	自复制自动机 zifuzhi zidongji
1326	self-tuning controller	régulateur m auto-ajustable	selbsteinstellender Regler	самонастраивающийся регулятор	controllore autosintonizzantesi	controlador m autoajustable	自己調整制御装置 jiko-chōsei-seigyo-sōchi	自校正控制 zijiaozheng kongzhi

No.	English	Français	Deutsch	Русский	Italiano	Español	日本語	中文
1327	self-tuning regulator	régulateur m auto-ajustable	selbsteinstellender Regulator (od. Konstanthalter)	самонастраивающийся регулятор	regolatore autosintonizzantesi	regulador m autoajustable	自己調整レギュレータ jiko-chōsei-regurēta	自校正调节器 zijiaozheng tiaojieqi
1328	semigraphic display	affichage m semi-graphique	halbgrafische Anzeige	полуграфический дисплей	visualizzatore semigrafico	presentación f semigráfica, visualización f semigráfica		半图解显示 bantujie xianshi
1329	sensing element	capteur m	Sensorelement	датчик, сенсор, чувствительный элемент	sonda	elemento m sensor, elemento m captador	検出要素 kenshutsu-yōso	敏感元件 mingan yuanjian
1330	sensitive	sensible	sensitiv, empfindlich	чувствительный	sensibile	sensible	敏感な binkan-na	灵敏的, 敏感的 lingminde, min'gande
1331	sensitivity	sensibilité f	Sensitivität, Empfindlichkeit	чувствительность	sensibilità, sensitività	sensibilidad f	感度 kando	灵敏度 lingmindu
1332	sensor	capteur m, transducteur m de mesure	Sensor, (Meß)fühler, Signalgeber	датчик, сенсор, чувствительный элемент	sensore	sensor m, captador m	検出器 sensa	传感器 chuanganqi
1333	separability	séparabilité f	Separabilität	сепарабельность	separabilità	separabilidad f	分離可能性 bunri-kanōsei	可分性 kefenxing
1334	sequential control	commande f séquentielle	Ablaufsteuerung	последовательное управление (регулирование), программное управление	controllo sequenziale	control m secuencial	シーケンス制御 shīkensu-seigyo	顺序控制 shunxu kongzhi
1335	sequential control algorithm	algorithme m de commande séquentielle	sequentieller Steueralgorithmus	последовательный алгоритм управления	algoritmo di controllo sequenziale	algoritmo m de control secuencial	逐次的制御アルゴリズム tsuijiteki-seigyo-arugorizumu	顺序控制算法 shunxu kongzhi suanfa
1336	sequential process-oriented control	régulation f orientée vers les procédés séquentiels	sequentielle prozeßorientierte Steuerung	последовательное процесс-ориентированное управление	controllo sequenziale guidato dal processo	control secuencial orientado al proceso	逐次過程的順序制御 tsuiji-kansōsa	面向过程的顺序控制 mianxiang guochengde shunxu kongzhi
1337	sequential return difference	différence f de retour séquentiel	sequentielle Rückführdifferenz	обратная разность последовательности	differenza di ritorno sequenziale	diferencia de retorno secuencial	逐次逆差 tsuiji-kansōsa	顺序回差 shunxu huicha
1338	sequential switching	commutation f séquentielle	sequentielles Schalten, Fortschalten	последовательное переключение	commutazione sequenziale	conmutación f secuencial	逐次切り換え tsuiji-kirikae	顺序开关 shunxu kaiguan
1339	sequential time-oriented control	régulation f orientée vers les procédés à temps séquentiel	zeitorientierte Ablaufsteuerung	последовательное время-ориентированное управление	controllo sequenziale guidato dal tempo	control secuencial orientado al tiempo		面向时间的顺序控制 mianxiang shijiande shunxu kongzhi

	English	French	German	Russian	Italian	Spanish	Japanese	Chinese
1340	series compensation	compensation f séries	Serien- od. Reihenkompensation	последовательная компенсация	compensazione in serie	compensación serie	直列補償 chokuretsu-hoshō	串联补偿 chuanlian buchang
1341	series transducer	transducteur m série	Transduktor in Reihenschaltung	последовательный магнитный усилитель	trasduttore in serie	transductor m serie	直列トランスダクタ chokuretsu-toransudakuta	串联饱和电抗器 chuanlian baohe diankangqi
1342	servicing system	système m d'entretien	Wartungssystem	система обслуживания	sistema di servizio	sistema m de servicio		服务系统 fuwu xitong
1343	servo system	système m asservi	Servosystem, Folgeregelsystem	сервосистема, следящая система	sistema ad asservimento	servosistema m	サーボ系 sābo-kei	伺服系统，随动系统 sifu xitong, suidong xitong
1344	servomechanism	servomécanisme m	Stellantrieb	сервомеханизм	servomeccanismo	servomecanismo m	サーボ機構 sābo-kikō	伺服机构 sifu jigou
1345	servomotor	servomoteur m	Stellmotor, Servomotor	сервомотор	servomotore	servomotor m	サーボモータ sābo-mōta	伺服马达，伺服电机 sifu mada, sifu dianji
1346	servomotor actuator	actionneur m à servomoteur	servobetätigter	сервомоторный исполнительный механизм, сервомоторный привод	attuatore a servomotore	actuador m servoasistido	サーボモータアクチュエータ sābomōta-akuchuēta	伺服马达驱动器 sifu mada qudongqi
1347	set (v)	afficher (v)	(ein) stellen, vorgeben	(предварительно) устанавливать	fissare, porre	ajustar, fijar, armar	設定する settei-suru	设定（动词）sheding (dongci)
1348	set point	consigne f, signal m de consigne	Sollwert, eingestellter Wert der Führungsgröße	точка задания, заданное значение задающее устройство	riferimento	valor m de consigna	設定値 setteichi	设定点 shedingdian
1349	set-point adjuster	afficheur m de consigne	Sollwertgeber, Sollwerteinsteller	супервизорное управление, управление заданием (=уставкой) регулятора	correttore del riferimento	ajustador m del valor de consigna	目標値調整器 mokuhyōti-chōseiki	设定点调整器 shedingdian tiaozhengqi
1350	set-point control	régulation f de maintien	Regelung mit Sollwertvorgabe, Sollwertführung, Führungssteuerung	точка задания, заданное значение	controllo sul riferimento	control m del valor de consigna	目標値制御 mokuhyōchi-seigyo	设定点控制 shedingdian kongzhi
1351	set value	valeur f de consigne	Einstellwert, Festwert	задающее устройство	valore di riferimento	valor m de consigna	設定値 mokuhyō-chi (hensū)	设定值 shedingzhi
1352	set value adjuster	afficheur m de consigne	Sollwerteinsteller	настройка, установка	correttore del valore di riferimento	ajustador m del valor de consigna	設定値調整器 setteichi-chōseiki	设定值调整器 shedingzhi tiaozhengqi

No.	English	French	German	Russian	Italian	Spanish	日本語	中文
1353	setting	affichage m	Einstellung, Einstellwert		impostazione	ajuste m, calado m	設定 settei	设定 sheding
1354	settling time	durée f de réglage, de rétablissement	Ausregelzeit, Einschwingzeit, Beruhigungszeit	время установления	tempo di assestamento	tiempo m de establecimiento	整定時間 seitei-jikan	过渡过程时间 guodu guocheng shijian
1355	shaping network	réseau m correcteur	Impulsformernetzwerk	формирующая цепь	rete formante	red f correctora, red f conformadora	整形回路 seikei-kairo	整形网络 zhengxing wangluo
1356	sharpness of resonance	acuité f de résonance	Resonanzüberhöhung	острота резонанса	acutezza (selettività) della risonanza	agudeza f de resonancia	共振の鋭さ kyōshin-no-surudosa	谐振锐度 xiezhen ruidu
1357	shift register	registre m à décalage	Schieberegister	регистр сдвига	registro di scorrimento	registro * de desplazamiento	シフトレジスタ shifuto-regisuta	移位寄存器 yiwei jicunqi
1358	shunt compensation	compensation f parallèle, shunt	Querkompensation, Parallelkompensation (elektr. Netz)	компенсация шунтированием, параллельная	compensazione a scambio	compensacion * paralelo	並列補償 heiretsu-hoshō	并联补偿 binglian buchang
1359	signal	signal m	Signal	сигнал	segnale	señal f	信号 shingō	信号 xinhao
1360	signal circuit	circuit m d'information	Signal(strom)kreis, Steuerstromkreis	сигнальная цепь	circuito di segnale (per l'elaborazione di segnali)	circuito m de señal	信号回路 shingō-kairo	信号线路，信号电路 xinhao xianlu, xinhao dianlu
1361	signal converter	convertisseur m de signal	Signalwandler	преобразователь сигналов	convertitore di segnali	convertidor m de señal	信号変換器 shingō-henkanki	信号转换器 xinhao zhuanhuanqi
1362	signal detection	détection f de signal	Signalerkennung	детектирование, обнаружение сигнала	rivelatore di segnali	detección * de señales	信号検出 shingō-kenshutsu	信号检测 xinhao jiance
1363	signal duration	durée f d'impulsion	Signaldauer	длительность сигнала	durata del segnale	duración f de señal	信号継続時間 shingō-keizoku-jikan	信号持续时间 xinhao chixu shijian
1364	signal flow diagram	graphe m de fluence des signaux	Signalflußdiagramm, Wirkplan	сигнальный граф	diagramma di flusso (di segnale)	diagrama m de fluencia (de las señales)	信号伝達線図 shingō-dentatsu-senzu	信号流图 xinhao liutu
1365	signal level	amplitude f du signal	Signalpegel	уровень сигнала	livello del segnale	nivel m de señal	信号レベル shingō-reberu	信号电平 xinhao dianping

No.	English	French	Russian	German	Italian	Spanish	日本語	中文
1366	signal processing	traitement m du signal	обработка сигнала	Signalverarbeitung	elaborazione di segnali	procesamiento m de señales	信号処理 shingō-shori	信号处理 xinhao chuli
1367	signal selector	sélecteur m de signaux	селектор сигналов	Signalwähler, Signalwahlschalter	selettore di segnale	selector m de señal	信号選択器 shingō-sentakuki	信号选择器 xinhao xuanzeqi
1368	signal synthesis	synthèse f de signal	синтез сигнала	Signalsynthese	sintesi di segnale	síntesis * de señales		信号综合 xinhao zonghe
1369	signal-to-noise ratio	rapport m signal sur bruit	отношение сигнал/шум	Signal-Rausch-Verhältnis, Signal-Störabstand	rapporto segnale-rumore	relación f señal/ruido	信号～雑音比 shingō-zatsuon-hi	信噪比 xinzaobi
1370	signature analysis	analyse f de signatures	сигнатурный анализ	Bit-Kombinationsanalyse (zur Identifizierung von Logikfehlern)	analisi di configurazioni caratteristiche	análisis de firmas		特征分析 tezhengfenxi
1371	simulation	simulation f	имитация, моделирование	Simulation	simulazione	simulación f	シミュレーション shimyurēshon	仿真 fangzhen
1372	simulation language	langage m de simulation	язык моделирования	Simulationssprache	linguaggio di simulazione	simulación * señales del lenguaje	シミュレーション言語 shimyurēshon-gengo	仿真语言 fangzhen yuyan
1373	simulator	simulateur m	моделирующее устройство	Simulator	simulatore	simulador m	シミュレータ shimyurēta	仿真器 fangzhenqi
1374	sine-cosine potentiometer	potentiomètre m à sinus-cosinus	синус-косинусный потенциометр	Sinus/Kosinuspotentiometer, Funktionspotentiometer	potenziometro sinu-cosinusoidale	potenciómetro m seno-coseno	正弦～余弦ポテンショメータ seigen-yogen-potensho-mēta	正弦余弦电位器 zhengxian yuxian dianweiqi
1375	sine wave	onde f sinusoïdale	синусоидальный сигнал	Sinuswelle, Sinusschwingung	onda sinusoidale	onda f sinusoidal, sinusoide m	正弦波 seigen-ha	正弦波 zhengxianbo
1376	single-input/single-output (SISO) system	système m à grandeurs d'entrée et de sortie uniques	система с одним входом-выходом	System mit einer Ein- u. Ausgangsgröße, SISO-System	sistema a un ingresso e un'uscita (SISO)	sistema m con una sola variable de entrada y de salida	1入力1出力系（システム）ichi-nyuryoku/ichi-shutsuryoku-kei (-shisutemu)	单输入单输出系统 danshuru danshuchu xitong
1377	single-speed floating action	action f flottante à vitesse unique	односкоростное астатическое воздействие	astatisches o. integrales (I-) Eingrößensystem	azione flottante a velocità unica	acción plotante de una velocidad	単速度浮動動作 tansokudo-fudō-dōsa	单速浮动作用 dansu fudong zuoyong
1378	single-variable system	système m à grandeurs d'entrée et de sortie uniques	скалярная система	Eingroßensystem	sistema mono-variable	sistema m monovariable	1変数系（システム）ichi-hensū-kei (-shisutemu)	单变量系统 danbianliang xitong

No.	English	French	German	Russian	Italian	Spanish	Japanese	Chinese
1379	singular point	point m singulier	singulärer Punkt	особая точка	punto singolare	punto m singular	特異点 tokui-ten	奇异点 qiyidian
1380	situation analyser	analyseur m de situation	Lageanalysator	анализатор ситуаций	analizzatore del contesto, o della situazione	analizador m de situación		状态分析器 zhuangtai fenxiqi
1381	situation control	régulation f de situation	Lageregelung (Steuerung)	ситуационное управление	controllo del contesto, o della situazione	control m de situación		状态控制 zhuangtai kongzhi
1382	situation recognition	reconnaissance f de situation	Lageerkennung	распознавание ситуаций	riconoscimento del contesto, o della situazione	reconocimiento m de situación		状态识别 zhuangtai shibie
1383	slave	esclave	Slave (Bussystem)	ведомая (станция), непривилегированный (режим)	subordinato	servocontrolar		从动的 congdongde
1384	slave (v)	asservir (v)	führen, nachführen	подчинять	subordinare	servocontrolar	従動させる jūdō-saseru	从动,役使 (动词) congdong, yishi (dongci)
1385	slave station	poste m esclave	Nebenstelle, Auswertestelle	ведомая станция	stazione subordinata	estación f esclava		从站 congzhan
1386	slide valve	vanne f à tiroir	Absperrschieber, Schieber(ventil)	золотник	valvola a slitta	válvula f de corredera	スライド弁 suraido-ben	滑阀,分流阀 huafa, fenliufa
1387	smooth (v)	lisser (v)	glätten, filtern, ebnen, ausgleichen	сглаживать	lisciare, interpolare, regolarizzare	alisar	平滑にする heikatsuka-suru	使平滑 (动词) shipinghua (dongci)
1388	solenoid actuator	actionneur m à solénoïde	Magnetstellantrieb, Magnetantrieb	соленоидный привод	attuatore a solenoide	actuador m de solenoide	ソレノイドアクチュエータ sorenoido-akuchueta	螺线管式驱动器 luoxianguanshi qudongqi
1389	source language	langage m source	Quellsprache, Primärsprache	исходный язык	linguaggio sorgente	lenguaje m fuente	ソースランゲージ (原始言語) sōsu-rangeiji (genshi-gengo)	源语言 yuanyuyan
1390	span	différence f des portées	Spanne, Bereich, Meßspanne	диапазон коррекции	lunghezza, ampiezza, estensione, copertura	intervalo (lapso)	スパン supan	跨距,变化范围 kuaju, bianhua fanwei
1391	specification	spécification f	Spezifikation, Vorschrift, Beschreibung	спецификация	specificazione	especificación f	仕様 shiyō	技术要求 jishu yaoqiu

No.	English	French	German	Russian	Italian	Spanish	Japanese	Chinese
1392	spectral density	densité f spectrale	Spektraldichte	спектральная плотность	densità spettrale	densidad f espectral	スペクトル密度 supekutoru-mitsudo	谱密度 pumidu
1393	spectral density function	fonction f de densité spectrale	spektrale Leistungsdichtefunktion	функция спектральной плотности	funzione densità spettrale	función de densidad espectral	スペクトル密度関数 supekutoru-mitsudo-kansū	谱密度函数 pumidu hanshu
1394	spectral/spectrum analyser	analyseur m harmonique, spectral	Spektralanalysator, Spektrumsanalysator	спектральный анализатор, анализатор спектра	analizzatore spettrale, o di spettro	analizador * de espectros	スペクトルアナライザ supekutoru-anaraiza	频谱分析仪 pinpu fenxiyi
1395	spectral/spectrum filter	filtre m harmonique, spectral	Farbfilter, Spektralfilter	спектральный фильтр	filtro spettrale	filtro * espectral	スペクトルフィルタ	频谱滤波器 pinpu lüboqi
1396	speech recognition	reconnaissance f de la parole	Spracherkennung	распознавание речи	riconoscimento del parlato	reconocimiento m de voz	音声認識 onsei-ninshiki	话音识别 yuyin shibie
1397	speed	vitesse f	Geschwindigkeit, Drehzahl	скорость	velocità	velocidad f	速度 sokudo	速率 sulü
1398	spherical resolver	résolveur m à bille	sphärischer Koordinatenwandler, Kugelkoordinatenwandler	шаровое решающее устройство	generatore sferico seno-coseno	resolver esférico	球座レゾルバ kyūkei-rezoruba	球形解算器 quixing jiesuanqi
1399	spike	pic m, pointe f	Spike, (Impuls-)Nadel, Fehlimpuls	пик	punta, impulso	pico m	スパイク supiaku	尖峰信号, 测试信号 jianfeng xinhao, ceshi xinhao
1400	spin axis	axe m de rotation	Drehachse	ось вращения	asse di rotazione	eje m de giro	スピン軸 supin-jiku	自旋轴 zixuanzhou
1401	spline-based filter	filtre m linguiforme	Splinegestütztes Filter	сплайновый фильтр	filtro a funzioni regolari di raccordo		スプライン型フィルタ supurain-gata-firuta	基于样条的滤波器 jiyu yangtiaode lüboqi
1402	split-field motor	moteur m à excitations inverses	Doppelfeldmaschine	двигатель с расщеплёнными полюсами	motore con avvolgimento ausiliario	motor m de campo partido	分割界磁電動機 bunkatsu-bunken-kaiji-dendōki	裂场电动机 liechang diandongji
1403	split-series motor	moteur m série à excitations inverses	Reihenmotor mit geteiltem Feld	двигатель постоянного тока с расцепленной сериесной обмоткой	motore ad avvolgimento ausiliario in serie	motor m serif de devanado partido	分割直巻界磁電動機 bunkatsu-chokken-kaiji-dendōki	裂绕串联电动机 liechang chuanji diandongji
1404	square matrix	matrice f carrée	quadratische Matrix	квадратная матрица	matrice quadrata	matriz * cuadrada	正方行列 seihō-gyōretsu	方阵 fangzhen

No.	English	French	German	Russian	Italian	Spanish	Japanese	Chinese
1405	square wave	onde f carrée	Rechteckwelle, Rechteckschwingung	прямоугольный сигнал	onda quadra	onda f cuadrada	方形波 hōkei-ha	方波 fangbo
1406	stability	stabilité f	Stabilität	устойчивость	stabilità	estabilidad f	安定性 anteisei	稳定性 wendingxing
1407	stability domain	domaine m de stabilité	Stabilitätsbereich	область устойчивости	regione di stabilità	campo m de estabilidad	安定領域 antei-ryōiki	稳定域 wendingyu
1408	stability limit	limite f de stabilité	Stabilitätsgrenze	предел устойчивости, граница устойчивости	limite di stabilità	limite m de estabilidad	安定限界 antei-genkai	稳定极限 wending jixian
1409	stabilizability	stabilisabilité f	Stabilisierbarkeit	стабилизируемость	stabilizzabilità	estabilizabilidad f	安定化可能性 anteika-kanōsei	可稳性 kewenxing
1410	stabilization	stabilisation f	Stabilisierung	стабилизация	stabilizzazione	estabilización f	安定化 anteika	镇定, 稳定 zhending, wending
1411	stabilize (v)	stabiliser (v)	stabilisieren	стабилизировать	stabilizzare	estabilizar	安定化する anteika-suru	使稳定 (动词) 镇定 shiwending (dongci)
1412	stabilizer	stabilisateur m	Konstanthalter	стабилизатор	stabilizzatore	estabilizador m	安定器 anteiki	稳定器 wendingqi
1413	stabilizing feedback	réaction f stabilisatrice	stabilisierende Rückführung	стабилизирующая обратная связь	retroazione stabilizzante	realimentación f estabilizadora	安定化フィードバック anteika-fidobakku	稳定 (镇定) 反馈 wending fankui
1414	stabilizing feedforward	action f directe stabilisatrice	stabilisierendes Vorwärtsglied	стабилизирующая прямая связь	azione stabilizzante diretta (ad anello aperto)	acción f anticipante estabilizadora	安定化フィードフォワード anteika-fidofowādo	稳定 (镇定) 前馈 wending qiankui
1415	stabilizing network	réseau m stabilisateur	stabilisierendes Netzwerk	стабилизирующая цепь	rete stabilizzatrice	red f estabilizadora	安定化回路 anteika-kairo	稳定 (镇定) 网络 wending wangluo
1416	stable	stable	stabil	устойчивый	stabile	estable	安定な antei-na	稳定的 wendingde
1417	stable state	état m stable	stabiler Zustand	устойчивое состояние	stato (di equilibrio) stabile	estado m estable	安定状態 antei-jōtai	稳定状态 wending zhuangtai

	English	French	German	Russian	Italian	Spanish	Japanese	Chinese
1418	standardized signal	signal *m* étalonné	Einheitssignal	стандартизованный сигнал	segnale normalizzato	señal *f* normalizada	標準化信号 hyōjunka-shingō	标准信号 biaozhun xinhao
1419	starting-pulse action	action *f* par impulsion initiale	Startimpuls-Wirkung	пусковое воздействие	(azione di) avviamento di un impulso	acción de pulso de arranque	起動パルス動作 kidō-parusu-dōsa	起动脉冲作用 qidong maichong zuoyong
1420	state	état *m*	Zustand, Zustandsgröße (Variable)	состояние	stato	estado *m*	状態 jōtai	状态 zhuangtai
1421	state estimation	estimation *f* d'état	Zustandsschätzung	оценивание состояния	stima dello stato	estimación *f* de estado	状態推定 jōtai-suitei	状态估计 zhuangtai guji
1422	state feedback	rétroaction *f* d'état	Zustandsrückführung	обратная связь по состоянию	retroazione dallo stato	realimentación *f* de estado	状態フィードバック jōtai-fidobakku	状态反馈 zhuangtai fankui
1423	state matrix	matrice *f* d'état	Zustandsmatrix	матрица состояний	matrice di stato	matriz de estado	状態行列 jōtai-gyōretsu	状态矩阵 zhuangtai juzhen
1424	state monitoring	supervision *f* d'état	Zustandsüberwachung	контроль состояния	sorveglianza sullo stato	supervisión *f* de estado	状態監視 jōtai-kanshi	状态监控 zhuangtai jiankong
1425	state observer	observateur *m* d'état	Zustandsbeobachter	наблюдатель состояния	osservatore dello stato	observador *m* de estado	状態観測器 jōtai-kansokuki	状态观测器 zhuangtai guanceqi
1426	state space	espace *m* d'état	Zustandsraum	пространство состояний	spazio di stato	espacio *m* de estado	状態空間 jōtai-kūkan	状态空间 zhuangtai kongjian
1427	state trajectory	trajectoire *f* d'état	Zustandstrajektorie	траектория состояния	traiettoria di stato	trayectoria *f* de estado	状態軌道 jōtai-kidō	状态轨迹 zhuangtai guiji
1428	state variable	variable *f* d'état	Zustandsvariable	переменная состояния	variabile di stato	variable *f* de estado	状態変数 jōtai-hensū	状态变量 zhuangtai bianliang
1429	state vector	vecteur *m* d'état	Zustandsvektor	вектор состояния	vettore di stato	vector *m* de estado	状態ベクトル jōtai-bekutoru	状态向量 zhuangtai xiangliang
1430	static	statique	statisch	статический	statico	estático	静的な seiteki-na	静态的 jingtaide

No.	English	French	German	Russian	Italian	Spanish	Japanese	Chinese
1431	static accuracy	précision f statique	statische Genauigkeit	статическая точность	precisione statica	precisión f estática	静的精度 seiteki-seido	静态精度 jingtai zhunquedu
1432	static controller	régulateur m statique	statischer Regler	статический регулятор	controllore non dinamico, o algebrico	controlador m estático, regulador m estático	定位制御器 teii-seigyo-ki	有静差控制器 youjingcha kongzhiqi
1433	static decoupling	découplage m statique	statische Endkopplung	статическая рассвязка	disaccoppiamento statico, o asintotico	desacoplo m estático	静的非干涉化 seiteki-hikanshōka	静态解耦 jingtai jie'ou
1434	static friction	frottement m solide	Ruhereibung	сухое трение	attrito statico	rozamiento m estático	固体摩擦 kotai-masatsu	静摩擦，干摩擦 jingmoca, ganmoca
1435	stationarity	stationnarité f	Stationarität	стационарность	stazionarietà	estacionariedad f	定常性 teijōsei	平稳性 pingwenxing
1436	status report	rapport m de statut	Statusbericht, Tätigkeits- u. Fehlerbericht	отчет (о состоянии)	rapporto sullo stato	informe m de situación	状態報告	状况报告 zhuangkuang baogao
1437	steady	établi	gleichbleibend, stationär	установившийся	permanente, costante	estacionario, permanente	定常な teijō-na	稳态的 wentaide
1438	steady oscillation	oscillation f entretenue, oscillation f permanente	Dauerschwingung	устойчивые колебания	oscillazione permanente	oscilación f permanente	定常振動 teijō-shindō	稳态振荡 wentai zhendang
1439	steady state	régime m établi, état m permanent	stationärer Zustand, Beharrungszustand	установившееся состояние	stato di regime	régimen m estacionario, régimen m permanente	定常状態 teijō-jōtai	稳态 wentai
1440	steady-state deviation	écart m, erreur f de statisme	bleibende (od. statische) Abweichung	установившееся отклонение	deviazione a regime	desviación f en régimen permanente	定常偏差 teijō-hensa	稳态偏差 wentai piancha
1441	steady-state error	erreur f de statisme	bleibender Fehler, bleibende (Regel)abweichung	установившаяся ошибка	errore a regime	error * en régimen permanente	定常誤差 teijō-hensa	稳态误差 wentai wucha
1442	steady-state value	valeur f en régime établi	eingeschwungener Zustand, Dauerwert	установившееся значение величины	valore a regime	valor m en régimen permanente	定常値 teijō-chi	稳态值 wentaizhi
1443	steepest descent	lieu m de plus grande pente	steilster Abfall (od. Anstieg)	наискорейший спуск	massima pendenza	máximo m descenso	最大傾斜法 saidaikeishahō	最速下降 zuisu xiajiang

	English	French	German	Russian	Italian	Spanish	Japanese	Chinese
1444	step-by-step action	action f par échelons	Schrittschaltung, Fortschaltung, Stufenwirkung	шаговое воздействие	azione passo-a-passo	acción f paso a paso	ステップ作用 suteppu-sayō	步进作用 bujin zuoyong
1445	step-by-step control	commande f pas-à-pas	Schrittsteuerung	шаговое управление (регулирование)	controllo passo-a-passo	control m paso a paso	階段制御 kaidan-seigyo	步进控制 bujin kongzhi
1446	step-by-step transmitter	moteur m pas-à-pas, transmetteur m pas-à-pas	Stufengeber, Stufentransmitter	шаговый передатчик	trasmettitore passo-a-passo	transmisor m paso a paso	階動トランスミッタ kaidō-toransumitta	步进发送器 bujin biansongqi
1447	step-enabling condition	condition f permettant un pas	Weiterschaltbedingung	условие допускающее скачки	condizione che abilita il passo	condición habilitadora de escalón		起步条件 qibu tiaojian
1448	step function	fonction f d'Heaviside, en échelon (de position)	Sprungfunktion	ступенчатая функция	funzione scalino	función f escalón	ステップ関数 suteppu-kansū	阶跃函数 jieyue hanshu
1449	step function response	réponse f indicielle	Sprungantwort	переходная характеристика	risposta a scalino	respuesta f a un escalón	ステップ応答 suteppu-ōtō	阶跃响应 jieyue xiangying
1450	step input	échelon m d'entrée	Eingangssprung	ступенчатый вход	ingresso a scalino	entrada f en escalón	ステップ入力 suteppu-nyūryoku	阶跃输入 jieyue shuru
1451	stepless action	action f progressive	stufenlose Anregung	непрерывное воздействие	azione regolare (senza salti)	acción f progresiva		无级进作用 wubujin zuoyong
1452	stepping action	action f par impulsion	Fortschaltung, Weiterschaltung	импульсное воздействие	azione a scalini	acción f escalonada, acción f paso a paso	ステップ動作 suteppu-dōsa	步进作用 bujin zuoyong
1453	stepping controller	régulateur m par impulsion, régulateur m à action pas-à-pas	Schrittsteuereinheit	регулятор пошагового действия	controllore a passo	controlador m paso a paso	ステッピング制御装置 suteppingu-seigyo-sōchi	步进式控制器 bujinshi kongzhiqi
1454	stepping motor	moteur m pas-à-pas	Schrittmotor	шаговый двигатель	motore a passo	motor m paso a paso	ステップモータ suteppu-mōta	步进马达, 步进电机 bujin mada, bujin dianji
1455	stepping relay	relais m pas-à-pas	Schrittrelais	шаговое реле, шаговый искатель	'relay' a passi	relé m paso a paso	ステップリレー suteppu-rirē	步进继电器 bujin jidianqi
1456	stepping switch	relais m pas-à-pas	Stufenschalter	шаговое реле, шаговый искатель	commutatore a passi	conmutador m paso a paso	ステップスイッチ suteppu-suitthi	步进开关 bujin kaiguan

No.	English	French	German	Russian	Italian	Spanish	Japanese	Chinese
1457	stiffness	raideur f	Steifigkeit	жёсткость	durezza	rigidez f	スティフネス sutifunesu	刚性，刚度 gangxing, gangdu
1458	stimulus	stimulus m	Anregung, Erregung, Reiz, Triggerimpuls	стимул	stimolo	estímulo m	刺激 shigeki	激励 jili
1459	stochastic	stochastique	stochastisch	стохастический	stocastico	estocástico m	確率的な kakuritsuteki-na	随机的 suijide
1460	stochastic approximation	approximation f stochastique	stochastische Approximation	стохастическая аппроксимация	approssimazione stocastica	aproximación f estocástica	確率近似 kakuritsu-kinji	随机逼近 siji bijin
1461	stochastic automaton	automate m stochastique	stochastischer Automat	случайный автомат, вероятностный автомат	automa stocastico	autómata m estocástico	確率オートマトン kakuritsu-ōtomaton	随机自动机 suiji zidongji
1462	stochastic control	régulation f stochastique	stochastische Regelung	стохастическое управление	controllo stocastico	control m estocástico	確率制御 kakuritsu-seigyo	随机控制 suiji kongzhi
1463	stochastic programing	programmation f stochastique	stochastische Programmierung	стохастическое программирование	programmazione stocastica	programación f estocástica	確率的計画法 kakuritsuteki-keikaku-hō	随机规划 suiji guihua
1464	stochastic system	système m stochastique	stochastisches System	стохастическая система	sistema stocastico	sistema m estocástico	確率系(システム) kakuritsu-kei (-shishutemu)	随机系统 suiji xitong
1465	stochastic variable	variable f stochastique	stochastische Größe (od. Variable)	стохастическая переменная	variabile stocastica	variable f estocástica	確率変数 kakuritsu-hensu	随机变量 suiji bianliang
1466	stop-band	bande f de coupure	Sperrband, Sperrbereich	полоса среза, полоса заграждаемых частот	arresta-banda	banda f de corte	阻止帯域 soshi-taiiki	阻带 zudai
1467	stop-band filter	filtre m à bande atténuée, à élimination de bande, coupe bande	Sperrbandfilter	полосно-заграждающий фильтр	filtro arresta-banda	filtro * eliminador de banda	帯域消去フィルタ taiiki-shōkyo-firuta	阻带滤波器 zudai lüboqi
1468	storage	mémoire f	Speicherung	память	memoria, accumulo	almacenamiento m	記憶装置 kioku-sōchi	存储器 cunchuqi
1469	storage element	organe m de mémoire	Speicherelement	элемент памяти, запоминающий элемент	elemento di memoria	elemento m de memoria	記憶要素 kioku-yōso	存储元件 cunchu yuanjian

No.	English	French	German	Russian	Italian	Spanish	日本語	中文
1470	store	mémoire f	Speicher	запоминающее устройство, память	magazzino	almacen m, memoria f	記憶 kioku	存储(器) cunchu(qi)
1471	structural constraint	limitation f structurelle	strukturelle Begrenzung (od. Einschränkung)	структурное ограничение	vincolo strutturale	restricción f estructural	構造約束 kōzōteki-kōsoku	结构约束 jiegou yueshu
1472	structural parameter	paramètre m structurel	Strukturparameter	структурный параметр	parametro strutturale	parámetro m estructural	構造的パラメータ kōzōteki-paraméta	结构参量 jiegou canliang
1473	structural stability	stabilité f structurelle	strukturelle Stabilität	структурная устойчивость	stabilità strutturale	estabilidad f estructural	構造安定性 kōzō-anteisei	结构稳定性 jiegou wendingxing
1474	structured language	langage m structuré	strukturierte (Programmier) Sprache	структурированный язык	linguaggio strutturato	lenguaje * estructurado	構造化言語 kōzōka-gengo	单位脉冲 jiegou yuyan
1475	suboptimal control	régulation f sous-optimale	suboptimale (od. teiloptimale) Regelung (Steuerung)	субоптимальное управление	controllo sub-ottimale	control m subóptimo	準最適制御 jun-saiteki-seigyo	次优控制 ciyoukongzhi
1476	suboptimal system	système m système sous-optimal	suboptimales System	субоптимальная система	sistema sub-ottimale	sistema m subóptimo	準最適系（システム） junsaiteki-kei (-shisutemu)	次优系统 ciyouxitong
1477	subsidence ratio	rapport m d'amortissement	Dämpfungsfaktor	коэффициент затухания	rapporto di sussidenza	relación de hundimiento	衰減比 shuaijianbi	衰减比 shuaijianbi
1478	subsidiary feedback	réaction f secondaire	Zusatzrückführung	вспомогательная обратная связь	retroazione sussidiaria	realimentación f secundaria	補助フィードバック hojo-fidobakku	辅助反馈 fuzhu fankui
1479	subspace	sous-espace m	Teilraum, Unterraum	подпространство	sottospazio	sub-espacio m	部分空間 bubun-kūkan	子空间 zikongjian
1480	summation element	additionneur m	Summationsglied, Summierwerk	сумматор, суммирующий элемент	elemento di somma	elemento m de suma	加算要素 kasan-yōso	求和元件 qouhe yuanjian
1481	summing amplifier	amplificateur m sommateur	Summierverstärker	суммирующий усилитель	amplificatore additivo	amplificador m sumador	加算増幅器 kasan-zōfukuki	加法放大器 jiafa fangdaqi
1482	summing element	sommateur m	Summierglied	сумматор, суммирующий элемент	elemento sommatore	(elemento) sumador m	加算要素 kasan-yōso	加法元件 jiafa yuanjian

No.	English	French	German	Russian	Italian	Spanish	Japanese	Chinese
1483	summing gear	engrenage m sommateur	Additonsgetriebe	суммирующая передача	ingranaggio sommatore	engranaje sumador	加算歯車 kasan-haguruma	求和机构 quohe jigou
1484	summing point	point m de sommation	Summierstelle	точка суммирования	punto di somma	punto m de suma	加え合わせ点 kuwae-awase-ten	相加点 xiangjiadian
1485	supervision	supervision f	Überwachung	контроль	supervisione	supervisión f	総括管理 sōkatsu-kanri	监督 jianguan
1486	supervisory control	régulation f de supervision	Überwachungssteuerung, Fernwirktechnik	супервизорное управление	controllo di supervisione	control m supervisor	管理制御 kanri-seigyo	监控 jiankong
1487	supply winding	enroulement m d'alimentation	Eingangswicklung, Speisewicklung	питающая обмотка, входная силовая обмотка	avvolgimento di alimentazione	arrollamiento m de alimentación, devanado n de alimentación	供給巻線 kyōkyū-makisen	电源绕组 dianyuan raozu
1488	swashplate pump	pompe f à plateau de réglage	Taumelscheibenpumpe		pompa rotativa a disco obliquo	bomba f de plato oscilante	斜板ポンプ shaban-ponpu	旋转斜盘泵 xuanzhuan xiepanbeng
1489	switch	interrupteur m, commutateur m	Schalter	переключатель	commutatore	interruptor m, conmutador m	スイッチ suicchi	开关 kaiguan
1490	switching algebra	algèbre f de branchement, algèbre m de commutation	Schaltalgebra	релейная алгебра	algebra di commutazione	álgebra m de conmutación	切り換え代数 kirikae-daisū	开关代数 kaiguan daishu
1491	switching function	fonction f de commutation	Schaltfunktion	переключающая функция	funzione di commutazione	función f de conmutación	スイッチ関数 suicchi-kansū	开关函数 kaiguan hanshu
1492	switching surface	surface f de commutation	Schaltoberfläche, Schaltfläche	поверхность переключения	superficie di commutazione	superficie f de conmutación	切り換え面 kirikae-men	开关面 kaiguanmian
1493	switching time	durée f de commutation	Schaltzeit	время переключения	istante di commutazione	tiempo m de conmutación	切り換え時間 kirikae-jikan	开关时间 kaiguan shijian
1494	switching value	valeur f de commutation	Schaltwert, Schaltpunkt	значение переключения	valore di commutazione	valor m de conmutación	切り換え値 kirikae-chi	开关值 kaiguanzhi
1495	switching variable	variable f de commutation	Schaltvariable	переменная переключения	variabile di commutazione	variable f de conmutación	切り換え変数 kirikae-hensū	开关变量 kaiguan bianliang

No.	English	French	German	Russian	Italian	Spanish	Japanese	Chinese
1496	synchro	synchro-machine f	Drehmelder, Drehfeld(system)	сельсин	sincro	sincro m	シンクロ shinkuro	自同步机，自整角机 zitongbuji, zizhengjiaoji
1497	synchro angle	angle m de synchro-machine	Gleichlaufwinkel	угол оси сельсина	sincro deviazione angolare	ángulo m de sincro	シンクロ角 shinkuro-kaku	同步角 tongbujiao
1498	synchro control differential transmitter	synchro-transmetteur m différentiel de régulation	Steuerdifferentialgeber (Drehmelder)	дифференциальный сельсин-датчик	sincro trasmettitore differenziale	sincrotransmisor m de control diferencial	シンクロ制御差動発信器 shinkuro-seigyo-sadō-hasshinki	控制式差动自同步变送器 kongzhishi chadong zitongbu biansongqi
1499	synchro control receiver	synchro-récepteur m de régulation	Steuerempfänger (Drehmelder)	сельсин-приемник	sincro ricevitore	sincroreceptor m de control	シンクロ制御受信器 shinkuro-seigyo-jushinki	控制式自同步接收器 kongzhishi zitongbu jieshouqi
1500	synchro control transformer	synchro-transformateur m de régulation	Steuerumsetzer (Drehmelder)	сельсин-приемник	sincro trasformatore	sincrocomparador m, sincrotransformador m de control	シンクロ制御変圧器 shinkuro-seigyo-henatsuki	控制式自同步变压器 kongzhishi zitongbu bianyaqi
1501	synchro control transmitter	synchro-transmetteur m de commande	Steuergeber (Drehmelder)	управляющий сельсин-датчик	sincro trasmettitore	sincrotransmisor m	シンクロ制御発信器 shinkuro-seigyo-hasshinki	控制式自同步发送器 kongzhiqi zitongbu biansongqi
1502	synchro indicator	synchro-indicateur m	Synchronanzeiger, Synchronindikator	индикаторный сельсин	sincro indicatore	sincroindicador m	シンクロ指示器 shinkuro-shijiki	同步指示器 tongbu zhishiqi
1503	synchro resolver	synchro-résolveur m, synchro-trigonomètre m	Koordinatenwandler, Funktionsdrehmelder	вращающий трансформатор	sincro 'resolver' (trasduttore di posizione angolare)	sincroresolver	シンクロレゾルバ shinkuro-rezoruba	同步解算器 tongbu jiesuanqi
1504	synchro torque differential receiver	synchro-récepteur m différentiel de puissance	(Dreh)moment-differentialempfänger	дифференциальный сельсин-приемник вращающего момента	sincro ricevitore differenziale di coppia	sincroreceptor m diferencial de par	シンクロトルク差動受信器 shinkuro-toruku-sadō-jushinki	力矩式同步差分接收器 liiushi tongbu chafen jieshouqi
1505	synchro torque differential transmitter	synchro-transmetteur m différentiel	(Dreh)moment-differentialgeber	дифференциальный сельсин-датчик вращающего момента	sincro trasmettitore differenziale di coppia	sincrotransmisor m diferencial de par	シンクロトルク差動発信器 shinkuro-toruku-sadō-hasshinki	力矩式同步差分变送器 liiushi tongbu chafen biansongqi
1506	synchro torque receiver	synchro-récepteur m de puissance	(Dreh)momentempfänger	сельсин-приемник вращающего момента	sincro ricevitore di coppia	sincroreceptor m de par	シンクロトルク受信器 shinkuro-tokuru-jushinki	力矩式同步接收器 liiushi tongbu jieshouqi
1507	synchro torque transmitter	synchro-transmetteur m de couple	(Dreh)momentgeber	сельсин-датчик вращающего момента	sincro trasmettitore di coppia	sincrotransmisor m de par	シンクロトルク発信器 shinkuro-tokuru-hasshinki	力矩式同步发送器 liiushi tongbu biansongqi
1508	synchro transmitter	synchro-transmetteur m	Drehmeldegeber	сельсин-(пере)датчик	sincro trasmettitore	sincrotransmisor m	シンクロ発信器 shinkuro-hasshinki	自同步发送器 zitongbu biansongqi

No.	English	français	Deutsch	русский	italiano	español	日本語	中文
1509	system analysis	analyse f des systèmes	Systemanalyse	системный анализ	analisi dei sistemi	análisis * de sistemas	システム解析 shisutemu-kaiseki	系统分析 xitong fenxi
1510	system architecture	architecture f de systèmes	Systemaufbau, Systemarchitektur	архитектура системы	architettura del sistema	arquitectura f del sistema	システム構成 shisutemu-kōsei	系统结构 xitong jiegou
1511	system deviation	erreur f de système	Systemabweichung, Regelabweichung	отклонение системы	deviazione del sistema	desviación del sistema	制御偏差 seigyo-hensa	系统偏差 xitong piancha
1512	system engineering	systémique f	Systementwurf, Systemplanung, Systemtechnik	проектирование систем, системотехника	ingegneria dei sistemi	ingeniería * de sistemas	システム工学 shisutemu-kōgaku	系统工程 xitong gongcheng
1513	system failure	panne f de réseau	Systemausfall	отказ системы	guasto del sistema	fallo m del sistema	システム故障 shisutemu-koshō	系统故障 xitong guzhang
1514	system matrix	matrice f du système	Systemmatrix	матрица системы	matrice del sistema	matriz f del sistema	システム行列 shisutemu-gyōretsu	系统矩阵 xitong juzhen
1515	system order	ordre m du système	Systemordnung	порядок системы	ordine del sistema	orden m sel sistema	システム次数 shisutemu-jisū	系统阶数 xitong jieshu
1516	system sensitivity	sensibilité f du système	Systempfindlichkeit (od. Sensitivität)	чувствительность системы	sensitività del sistema	sensibilidad f del sistema	システム感度 shisutemu-kando	系统灵敏度 xitong lingmindu
1517	system synthesis	synthèse f de systèmes	Systemsynthese	синтез системы	sintesi del sistema	síntesis * de sistemas	システムシンセシス shisutemu-shinseshisu	系统综合 xitong zonghe
1518	system theory	théorie f des systèmes	Systemtheorie	теория систем	teoria dei sistemi	teoría * de sistemas	システム理論 shisutemu-riron	系统理论 xitong lilun
1519	system transfer function	transmittance f du système	Systemübertragunsfunktion	передаточная функция системы	funzione di trasferimento del sistema	función f de transferencia del sistema	システム伝達関数 shisutemu-dentatsu-kansū	系统传递函数 xitong chuandi hanshu
1520	tacho generator	génératrice f tachymétrique	Tachogenerator, Tachodynamo	тахогенератор	dinamo tachimetrica	tacho generator generador m tacométrico, generatriz f tacométrica	タコジェネレータ tako-jenerēta	测速电机 cesu dianji
1521	tachometer	génératrice f tachymétrique, tachymètre m	Tochmeter, Drehzahlmesser	тахометр	tachimetro	tacómetro m	タコメータ takomēta	转速表 zhuansubiao

No.	English	French	German	Russian	Italian	Spanish	Japanese	Chinese
1522	tapered potentiometer	potentiomètre m non-linéaire	nichtlineares Potentiometer	нелинейный потенциометр	potenziometro sagomato	potenciómetro m no lineal		非线性电位计 fexianxing dianweiji
1523	tapped potentiometer	potentiomètre m à prises	Potentiometer mit Anzapfung	потенциометр с отводами	potenziometro a prese intermedie	potenciómetro m no lineal	タップ付きポテンショメータ tappu-tsuki-potensho-mēta	抽头式电位计 choutoushi dianweiji
1524	target control	régulation f de suivi	Zielregelung	управление по цели	controllo mirato	control m de objetivo	ターゲット制御 tāgetto-seigyo	目标控制 mubiao kongzhi
1525	telecontrol	télécommande f	Fernbedienung, Fernwirken, Fernwirktechnik (engineering)	дистанционное управление, телеуправление	telecontrollo	telecontrol m	遠隔制御 enkaku-seigyo	遥控 yaokong
1526	telecommunication	télécommunication f	Telekommunikation, Nachrichtenübertragung	телекоммуникация	telecomunicazione	telecomunicación *	遠隔通信 enkaku-tsūshin	远程通信 yuancheng tongxin
1527	telemetry	télémesure f	Telemetrie	телеметрия	telemetria	telemedida *	テレメータリング teremētaringu	遥测 yaoce
1528	terminal control	régulation f en prise directe sur la procédé	Endwertregelung	терминальное управление	controllo terminale	control m terminal	終点制御 shūten-seigyo	终端控制 zhongduan kongzhi
1529	ternary logic	logique f ternaire	Dreistufenlogik	троичная логика	logica ternaria	lógica * ternaria	3値論理 sanchi-ronri	三进制逻辑 sanjinzhi luoji
1530	thermal noise	bruit m thermique	Wärmerauschen od. thermisches Rauschen	тепловой шум, тепловые помехи	rumore termico	ruido * térmico	熱雑音 netsu-zatsuon	热噪声 rezaosheng
1531	thermistor	thermistance f	Thermistor, temperaturabhängiger Widerstand, Heißleiter	термистор, терморезистор	termistore	termistor *	サーミスタ sāmisuta	热敏电阻 remindianzu
1532	three-step action	action f à trois niveaux	Dreipunktverhalten	трехпозиционное воздействие	azione a tre scalini	acción f a tres niveles	3値動作 sanchi-dōsa	三位作用 sanwei zuoyong
1533	three-term action	action f à trois termes	PID-Verhalten (drei Regelungsstrategien)	ПИД-воздействие	azione a tre termini	acción triple	3項動作 sankō-dōsa	PID三项作用 PID sanxiang zuoyong
1534	three-term control	régulation f à action triple	PID-Regelung (drei Regelungsstrategien)	ПИД-регулирование	controllo a tre termini	control m de acción triple	3項動作制御（ＰＩＤ）sankō-dōsa-seigyo	PID三项控制 PID sanxiang kongzhi

	English	French	German	Russian	Italian	Spanish	Japanese	Chinese
1535	three-term controller	régulateur m à action triple	PID-Regler, Regler mit drei dynamischen Strategien	ПИД-регулятор	controllore a tre termini	controlador m de acción triple	3項動作調節器 sankō-dōsa-chōsetsuki	PID三項調制器 PID sanxiang kongzhiqi
1536	threshold	seuil m	Ansprechgrenze, Schwellwert	порог	soglia	umbral m	いき値 ikichi	阈 yu
1537	threshold logic	seuil m logique	Schwellwertlogik	пороговая логика	logica a soglia	lógica * de umbral	域値論理 ikichi-ronri	阈逻辑 yuluoji
1538	threshold of resolution	seuil m de résolution	Schwellwert des Auflösungsvermögens	разрешающая способность	soglia di risoluzione	umbral m de resolución	分解のいき値 bunkai-no-ikichi	分辨阈 fenbianyu
1539	threshold value	valeur f seuil	Schwellwert, Schwellenwert	пороговое значение	valore di soglia	valor m umbral	いき値 ikichi	阈值 yuzhi
1540	time constant	constante f de temps	Zeitkonstante	постоянная времени	costante di tempo	constante f de tiempo	時定数 jiteisū	时间常数 shijian changshu
1541	time delay	retard m	Zeitverzögerung	задержка, запаздывание	ritardo di tempo	retardo m	時間おくれ jikan-okure	时延 shiyan
1542	time domain	domaine m temporel	Zeitbereich	временная область	dominio del tempo	campo m temporal	時間領域 jikan-ryōiki	时域 shiyu
1543	time-invariant system	système m permanent	zeitinvariantes System	стационарная система	sistema invariante nel tempo (o tempo-invariante)	sistema m invariante en el tiempo	時不変系(システム) jifuhen-kei (-shisutemu)	定常系统, 非时变系统 dingchang xitong, feishibian xitong
1544	time lag	retard m à la montée	Verzögerungszeit, zeitliche Nacheilung, zeitlicher Verzug	время задержки, время запаздывания	ritardo di tempo	retardo m inicial, tiempo m muerto	時間遅れ jikan-okure	时滞 shizhi
1545	time-optimal control	régulation f brachystochrone	zeitoptimale Regelung (Steuerung)	оптимальное по быстродействию управление	controllo in tempo minimo	control m de tiempo óptimo	時間最速制御 jikan-saiteki-seigyo	时间最优控制 shijian zuiyou kongzhi
1546	time response	temps m de réponse	Zeitverhalten, Zeitantwort, Ansprechzeit	временная характеристика	risposta (nel tempo)	respuesta f temporal	時間応答 jikan-ōtō	时间响应 shijian xiangying
1547	time schedule control	régulation f à programme	Zeitplanregelung (Steuerung)	программное управление (регулирование)	controllo a programmazione nel tempo	control m por consigna temporal	プログラム制御 puroguramu-seigyo	编程控制 biancheng kongzhi

No.	English	French	German	Russian	Italian	Spanish	Japanese	Chinese
1548	time schedule controller	régulateur m à programme	Zeitplanregler	программный регулятор	controllore a programmazione nel tempo	controlador m por consigna temporal	プログラム調節器 puroguramu-chōsetsuki	编程控制器 biancheng kongzhiqi
1549	time shared control	régulation f en temps partagé	Zeitmuliplexregelung (Steuerung)	управление с разделением времени	controllo a partizione del tempo ('time-sharing')	regulación f en tiempo compartido	時分割制御 jibunkatsu-seigyo	分时控制 fenshi kongzhi
1550	time-sharing program	programmation f en temps partagé	Time-Sharingprogramm, gleichzeitiges arbeiten mehrerer Programme	программа разделения времени	programma a partizione del tempo ('time-sharing')	programa m de tiempo compartido	時分割プログラム jibunkatsu-puroguramu	分时程序 fenshi chengxu
1551	time-sharing system	système m en temps partagé	Mehrbenutzersystem, Teilnehmer-(Reihen) System	система разделения времени	sistema a partizione del tempo ('time-sharing')	sistema m de tiempo compartido	時分割系 (システム) jibunkatsu-kei (-shisutemu)	分时系统 fenshi xitong
1552	time-varying system	système m évolutif	zeitvariables (od. zeitunabhängiges) System	нестационарная система	sistema variante nel tempo (o tempo-variante)	sistema m variable en el tiempo	時変系 (システム) jihen-kei (shisutemu)	时变系统 shibian xitong
1553	torque amplifier	cabestan m	Drehmomentverstärker	усилитель вращающего момента	amplificatore di coppia	amplificador de par	トルク増幅器 toruku-zōfukuki	力矩放大器 liju fangdaqi
1554	torque control	régulation f de couple	Drehmomentregelung	управление вращающим моментом	controllo di coppia	control * de par	トルク制御 toruku-seigyo	力矩控制 liju kongzhi
1555	torque motor	moteur m à couple constant	Drehmomentmotor	тормозной двигатель, моментный двигатель	motore coppia	motor m de par	トルクモータ toruku-mōta	力矩马达 liju mada
1556	torque synchro	synchro-machine f de puissance	Drehwinkelsynchro	сельсин-датчик вращающего момента	sincro coppia	sincro m de potencia	トルクシンクロ toruku-shinkuro	力矩式自同步机 lijushi zitongbuji
1557	torsional vibration damper	amortisseur m de vibration à torsion	Drehschwingungs- od. Torsionsschwingungs-dämpfer	гаситель крутильных колебаний	smorzatore di vibrazioni torsionali	amortiguador m torsional de vibraciones	ねじり振動減衰器 nejiri-shindō-gensuiki	扭振阻尼器 niuzhen zuniqi
1558	tracking system	système m de poursuite	Kursführungssystem, Nachlaufsystem	слеживающая система	sistema a inseguimento	sistema m de seguimiento	追随システム tsuiseki-shisutemu	跟踪系统 genzong xitong
1559	trajectory	trajectoire f	Trajektorie	траектория	traiettoria	trayectoria f	軌道 kidō	轨迹·滑道 guiji, dandao
1560	transcoder	transcodeur m	transienter Fehler	преобразователь кодов	transcodificatore	convertidor m de código, transcodificador m	トランスコーダ toransu-kōda	代码转换器 daima zhuanhuanqi

No.	English	French	German	Russian	Italian	Spanish	Japanese	Chinese
1561	transducer	transducteur m	Wandler, (Meßumformer)	преобразователь, датчик	trasduttore	transductor m	トランスデューサ toransudyūsa	变换器, 变送器 bianhuanqi, biansongqi
1562	transductor	transducteur m	Transduktor, Regeldrossel	трансдуктор, магнитный усилитель	trasduttore	reactor m saturable, transductor m magnético	トランスダクタ toransudakuta	饱和电抗器 baohe diankangqi
1563	transductor element	organe m de transducteur	Transduktorelement, Wandler	дроссельный магнитный усилитель	elemento trasduttore	elemento m transductor	トランスダクタ要素 toransudakuta-yōso	饱和电抗器元件 baohe diankangqi yuanjian
1564	transfer behaviour	comportement m de transfert	Übertragungsverhalten	свойства передаточного поведения	comportamento rispetto alla trasmissione (o trasferimento)	comportamiento de transferencia		转换行为, 传递行为 zhuanhuan xingwei, chuandi xingwei
1565	transfer contacts	contacts m de basculement	Umschaltkontakte	переключающие контакты	contatti di trasferimento	contactos m inversores		转换触点 zhuanhuan chudian
1566	transfer element	circuit m de transfert	Übertragungsglied	передаточный элемент	elemento di trasferimento	elemento de transferencia	伝達要素 dentatsu-yōso	转换元件 zhuanhuan yuanjian
1567	transfer function	transmittance f, fonction f de transfert	Übertragungsfunktion	передаточная функция	funzione di trasferimento	función f de transferencia, transmitancia f	伝達関数 dentatsu-kansū	传递函数 chuandi hanshu
1568	transfer function matrix	transmittance f matricielle	Übertragungsmatrix	матрица передаточной функции	matrice f (funzioni di) trasferimento	matriz f de función de transferencia	伝達関数行列 dentatsu-kansū-gyōretsu	传递函数矩阵 chuandi hanshu juzhen
1569	transfer locus	lieu m de transfert	Ortskurve des Frequenzganges	амплитудно-фазовая характеристика	diagramma di trasferimento	lugar m geométrico de transferencia	伝達軌跡 dentatsu-kiseki	传递轨迹 chuandi guiji
1570	transformer	transformateur m	Transformator	трансформатор	trasformatore	transformador m	変圧器 変成器 henatsuki henseiki	变压器 bianyaqi
1571	transient	transitoire	transienter (od. flüchtiger) Vorgang, Ausgleichsvorgang	переходный	transitorio	transitorio m	過渡的な katoteki-na	瞬态的 shuntaide
1572	transient deviation	écart m transitoire	vorübergehende Abweichung	динамическое отклонение, ошибка	deviazione transitoria	desviación f transitoria	過渡偏差 kato-hensa	瞬态偏差 shuntai piancha
1573	transient error	écart m, erreur f transitoire	transienter Fehler	переходная ошибка	errore transitorio	error * transitorio	過渡誤差 kato-hensa	瞬态误差 shuntai wucha

No.	English	français	Deutsch	русский	italiano	español	日本語	中文
1574	transient response	réponse f transitoire	Übergangsverhalten, Übergangsfunktion	переходная характеристика	risposta transitoria	respuesta f transitoria	過渡応答 kato-ōtō	瞬态响应 shuntai xiangying
1575	transient stability	stabilité f transitoire	Stabilität des Übergangsverhaltens	переходная устойчивость	stabilità transitoria	estabilidad f transitoria	過渡安定性 kato-anteisei	瞬态稳定性 shuntai wendingxing
1576	transient state	régime m transitoire	Einschwingzustand, transienter Zustand	переходное состояние	stato transitorio	régimen m transitorio	過渡状態 kato-jōtai	瞬态 shuntai
1577	transient system deviation	écart m de consigne transitoire	transiente Systemabweichung	динамическое отклонение системы	deviazione transitoria di sistema	desviación f transitoria (del valor de consigna)	過渡偏差 kato-hensa	系统瞬态偏差 xitong shuntai piancha
1578	transition matrix	matrice f de transition	Übergangsmatrix	переходная матрица	matrice di transizione	matriz f de transición	遷移行列 sen-i-gyōretsu	转移矩阵 zhuanyi juzhen
1579	transition time	temps m de lancer	Übergangszeit	время разгона, время перехода	istante di transizione	tiempo m de transición	遷移時間 sen-i-jikan	转移时间 zhuanyi shijian
1580	transmitter	transmetteur m	Transmitter	передатчик, датчик	trasmettitore	transmisor m	伝送器 densōki	变送器 biansongqi
1581	transport delay	retard m de transport	Transportverzögerung (Signal)	транспортное запаздывание	ritardo di trasporto	retardo m de recorrido	伝送遅れ densō-okure	传输延迟 chuanshu yanchi
1582	tree structure	arborescence f, structure f arborescente	Baumstruktur (Netzwerk)	древовидная структура	struttura ad albero	estructura f arborescente	木構造 ki-kōzō	树结构 shu jiegou
1583	trigger element	circuit m basculeur	Triggereinrichtung	триггер	elemento d'innesco	elemento m disparador	トリガ要素 toriga-yōso	触发元件 chufa yuanjian
1584	truth table	table f de définition	Wahrheitstabelle	таблица истинности	tabella della verità	tabla f de correspondencia, tabla f de verdad	真理値表 shinrichi-hyō	真值表 zhenzhibiao
1585	tuned torsional vibration damper	amortisseur m de vibration à torsion accordé	abgestimmte Torsionsschwingungsdämpfung	настроенный гаситель крутильных колебаний	smorzatore sintonizzato di vibrazioni torsionali	amortiguador m torsional sintonizado	共振形ねじり振動減衰器 kyōshingata-nejirishindō-gensuiki	调谐扭振阻尼器 tiaoxie niouzhen zuniqi
1586	Turing machine	machine f de Turing	Turing-Maschine	машина Тюринга	macchina di Turing	máquina * de Turing	チューリング機械 churingu-kikai	图灵机 tulingji

No.	English	French	German	Russian	Italian	Spanish	Japanese	Chinese
1587	turnkey system	système m à verrouillage	schlüsselfertiges System	система под ключ	sistema chiavi-in-mano	sistema m llaves en mano	ターンキーシステム tānki-shisutemu	交钥匙系统 jiaoyaoshi xitong
1588	two-phase induction motor	moteur m à induction diphasé	Zweiphasen-induktionsmotor	двухфазовый индукционный мотор	motore a induzione bifase	motor m de inducción bifásico	2相誘導電動機 nisō-yūdō-dendōki	两相异步电机 liangxiang yibu dianji
1589	two-speed controller	régulateur m à deux vitesses d'action	Zweilaufregler	двухскоростной регулятор	controllore a due velocità	regulador m de dos velocidades	2値調節器 nichi-chōsetsuki	双速控制器 shuangsu konzhiqi
1590	two-step action	action f à deux niveaux	Zweipunktverhalten	двухпозиционное воздействие	azione a due scalini	acción f a dos niveles	2値動作 nichi-dōsa	两位作用 liangwei zuoyong
1591	two-step controller	régulateur m à deux échelons	Zweipunktregler	двухпозиционный регулятор	controllore a due scalini	regulador m a dos niveles	2値調節器 nichi-chōsetsuki	二位控制器 erwei kongzhiqi
1592	two-step neutral zone action	action f à deux niveaux avec zone d'insensibilité	Zweipunktverhalten mit Überlappung	двухпозиционное воздействие с перерегулированием	azione a due scalini con zona morta	acción f a dos niveles con zona muerta	中立素付2値動作 chūritsutaitsuki-nichi-dōsa	中性区的两位作用 zhongxingqude liangwei zuoyong
1593	two-term action	action f à deux termes	PI- bzw. PD-Verhalten (zwei Regelungsstrategien)	ПД-воздействие, ПИ-воздействие	azione a due termini	acción doble	2項動作 nikō-dōsa	二项作用 erxiang zuoyong
1594	two-term action with overlap	action f à deux niveaux avec recouvrement	PI- bzw. PD-Verhalten mit Überlappung	ПД-воздействие с перерегулированием, ПИ-воздействие с перерегулированием	azione a due termini con sovrapposizione	acción doble con sobreposición	重合付2値動作 jūgōtsuki-nichi-dōsa	具有重叠的二项作用 juyou chongdiede erxiang zuoyong
1595	two-term control	régulation f à action double	PI- bzw. PD-Regelung (zwei Regelungsstrategien)	ПД-регулирование, ПИ-регулирование	controllo a due termini	control m bi-acción	2項動作制御 (PI、PD) nikō-dōsa-seigyo	二项控制 erxiang kongzhi
1596	two-term controller	régulateur m à action double	Regler mit zwei dynamischen Strategien (PD- bzw. PI-Verhalten)	ПД-регулятор, ПИ-регулятор	controllore a due termini	controlador con dos acciónes	二項調節器 nikō-chōsetsuki	二项控制器 erxiang kongzhiqi
1597	ultimately controlled variable	grandeur f réglée finale	nichtsteuerbar	предельная регулируемая величина, конечная регулируемая величина	variabile controllata terminale	magnitud f controlada final		最终被控变量 zuizhong beikong bianliang
1598	uncontrollable	ingouvernabilité f, incommandabilité f	ungedämpfte Schwingung	неуправляемый	non controllabile	incontrollable	不可制御な fuka-seigyo-na	不可控的 bukekongde
1599	undamped frequency	fréquence f naturelle	unterkritische Dämpfung	собственная частота	frequenza (o pulsazione) non smorzata	frecuencia f propia	非減衰周波数 hi-gensui-shūhasū	无阻尼频率 wuzuni pinlü

No.	English	French	German	Russian	Italian	Spanish	Japanese	Chinese
1600	underdamping	amortissement m sous critique	unterkritische Dämpfung	недостаточное демпфирование, слабое затухание	subsmorzamento	amortiguamiento m subcrítico	不足減衰 fusoku-gensui	欠阻尼 qianzuni
1601	unit impulse	impulsion f unitaire	Einheitsimpuls	единичный импульс, дельта импульс	impulso unitario	impulso m unitario, impulso m de Dirac	単位インパルス tan-i-inparusu	单位脉冲 danwei chongji
1602	unit-step function	fonction f échelon unitaire	Einheitsprungfunktion	единичная (ступенчатая) функция	funzione scalino unitario	(función) escalón m unitario	単位ステップ関数 tan-i-suteppu-kansū	单位阶跃函数 danwei jieyue hanshu
1603	unit-step response	réponse f indicielle	Einheitssprungantwort, Übergangsfunktion	переходная характеристика	risposta a scalino unitario	respuesta f unitaria	単位ステップ応答 tan-i-suteppu-ōtō	单位阶跃响应 danwei jieyue xiangying
1604	unmonitored control system	système m de commande en boucle ouverte	Steuersystem	разомкнутая система управления	sistema di controllo senza sorveglianza	sistema m de mando en anillo abierto		非监控系统 feijiankong xitong
1605	unobservable	inobservabilité f	nichtbeobachtbar	ненаблюдаемый	non osservabile	inobservable	不可観測な fuka-kansoku-na	不可观测的 bukeguancede
1606	unstable	instable	instabil	неустойчивый	instabile	inestable	不安定な fuantei-na	不稳定的 buwendingde
1607	upward compatibility	compatibilité f avec l'amont	Kompatibilität, Verträglichkeit	совместимость (снизу) вверх	compatibilità verso l'alto	compatibilidad f ascendente	上位両立性 上位適合性 jōi-ryōritsusei jōi-tekigōsei	向上兼容 xiangshang jianrong
1608	value	valeur f	Meßwert, Zahlenwert	значение	valore	valor m	値 atai	值，数值 zhi, shuzhi
1609	valve	vanne f	Ventil, Absperr- oder Regelorgan	клапан, вентиль	valvola	válvula f	弁 ben	阀，活门 fa, huomen
1610	valve lap	recouvrement m de la vanne à tiroir	Kolbenweg	перекрытие вентиля	alvo (o alveo) della valvola	solapamiento m de válvula	弁の重合量 ben-no-jūgōryō	阀余面 fayumian
1611	valve plug	clapet m de vanne	Ventilkegel	конус клапана	spina della valvola	macho m de válvula	弁栓 bensen	阀塞 fasai
1612	valve port	lumière f de vanne	Ventilkanal	проходное отверстие клапана	bocca della valvola	lumbrera f de la válvula	弁ポート ben-pōto	阀口 fakou

	English	French	German	Russian	Italian	Spanish	Japanese	Chinese
1613	valve positioner	positionneur *m* de vanne	Ventilstellungsregler	клапанный позиционер	servo-posizionatore (abitualm. pneumatico) dello stelo	posicionador *m* de válvula	弁ポジショナ ben-pojishona	阀门定位器 famen dingweiqi
1614	vane	palette *f*	Flansch, (Dreh)flügel	крыло, лопатка (клапана)	pala	vano	ベーン bēn	叶片, 挡板 yepian, dangban
1615	vane pump	pompe *f* à ailettes	Flügelradpumpe	крыльчатый (лопастный) насос	pompa a palette	bomba *f* de paletas	ベーンポンプ ben-ponpu	叶轮泵 yelunbeng
1616	variable	grandeur *f*	Variable	переменная (величина) variabile	variabile	variable	変数 hensū	变量, 可变的 bianliang, kebiande
1617	variable command control (closed loop)	régulation *f* de correspondance, asservissement *m*	Führungsregelung	следящее управление (замкнутое)	controllo a comando variabile (ad anello chiuso)	control *m* seguimiento, servomando *m*	プログラム制御 puroguramu-seigyo	可变指令控制 (附环) kebian zhiling kongzhi (bihuan)
1618	variable-delivery pump	pompe *f* à débit variable	Pumpe mit veränderlicher Fördermenge	насос переменной производительности	pompa a portata variabile	bomba *f* de caudal variable	可変容量形ポンプ kahen-yōryōgata-ponpu	变量传送泵 bianliang chuansongbeng
1619	variance matrix	matrice *f* de variance	Kovarianzmatrix	дисперсионная матрица	matrice (di) varianza	matriz *f* de varianza	分散行列 bunsan-gyōretsu	方差矩阵 fangcha juzhen
1620	velocity	vitesse *f*	Geschwindigkeit	скорость	velocità	velocidad *f*	速度 sokudo	速度 sudu
1621	velocity error	erreur *f* de vitesse	Geschwindigkeit-sfehler	скоростная погрешность	errore di velocità	error *m* de velocidad	速度偏差 sokudo-hensa	速度误差 sudu wucha
1622	velocity feedback	réaction *f* tachymétrique	Geschwindigkeits-rückführung	обратная связь по скорости	retroazione di velocità	realimentación *f* taquimétrica	速度フィードバック sokudo-fidobakku	速度反馈 sudu fankui
1623	vibrating reed	lame *f* vibrante	schwingende Zunge, Zungenfrequenz-messer (frequency meter)	вибратор	linguetta vibrante	lámina *f* vibrante	振動片 shindō-hen	振簧 zhenhuang
1624	vibration damper	amortisseur *m* de vibration	Schwingungsdämpfer	виброгаситель	smorzatore di vibrazioni	amortiguador *m* de vibraciones	振動減衰器 shindō-gensuiki	振动阻尼器 zhendong zuniqi
1625	viscous damping	amortissement *m* visqueux	viskose Dämpfung	вязкое демпфирование	smorzamento viscoso	amortiguamiento *m* viscoso	粘性減衰 nensei-gensui	粘性阻尼 nianxing zuni

No.	English	French	German	Russian	Italian	Spanish	Japanese	Chinese
1626	viscous friction	frottement m visqueux	zähe (od. viskose) Reibung	вязкое трение	attrito viscoso	rozamiento m viscoso	粘性摩擦 nensei-masatsu	粘性摩擦 nianxing moca
1627	voltage amplifier	amplificateur m de tension	Spannungsverstärker	усилитель напряжения	amplificatore di tensione	amplificador m de corriente	電圧増幅器 denatsu-zōfukuki	电压放大器 dianya fangdaqi
1628	voltage stabilizer	stabilisateur m de tension	Spannungs-Konstanthalter, Spannungsstabilisator	стабилизатор напряжения	stabilizzatore di tensione	estabilizador m de tensión	電圧安定器 denatsu-anteiki	电压稳定器 dianya wendingqi
1629	Walsh function	fonction f de Walsh	Walsh-Funktion	функция Уолша	funzione di Walsh	Booleana * de Walsh	ウォルシュ関数 Walsh-kansū	沃尔什函数 weershi hanshu
1630	Ward-Leonard drive	groupe m de Ward-Léonard	Ward-Leonard-Antrieb	привод по схеме Леонарда, привод по схеме генератор-двигатель	gruppo Ward-Leonard	grupo Ward-Leonard m	ワード・レオナード駆動 Ward-Leonard-kudō	华特-利奥纳特传动机组 huate-li'aonate chuandong jizu
1631	wave	onde f	Welle, Schwingung (Impulsfolge)	сигнал	onda	onda f	波 nami	波 bo
1632	wave train	train m d'ondes	Wellenzug	последовательность сигналов	treno d'onde	tren m de ondas	波列 haretsu	波列 bolie
1633	waveform analysis	analyse f de formes d'ondes	Analyse der Wellenform od. der Form einer Impulsfolge	анализ (формы) сигнала	analisi di forma d'onda	análisis m de forma de ondas	波形解析 hakei-kaiseki	波形分析 boxing fenxi
1634	weighting function	fonction f de pondération	Gewichtsfunktion	весовая функция	funzione peso	función f de peso	重み関数 omomi-kansū	权函数 quanhanshu
1635	wheel-and-disc integrator	intégrateur m à disque et molette	Scheibenintegrierer	роликовый фрикционный интегратор	integratore a ruota-e-disco	integrador de rueda y disco	ローラ・円盤形積分器 rōra-enbangata-sekibunki	轮盘式积分器 lunpanshi jifenqi
1636	white noise	bruit m blanc	weißes Rauschen (od. Geräusch)	белый шум	rumore bianco	ruido m blanco	白色雑音 hakushoku-zatsuon	白噪声 baizaosheng
1637	Wiener filter	filtre m de Wiener	Wiener-Filter	фильтр Винера	filtro di Wiener	filtro m de Wiener	ウィーナーフィルタ Wiener-firuta	维纳滤波器 weina lüboqi
1638	wire-wound potentiometer	potentiomètre m bobiné	Drahtpotentiometer	проволочный потенциометр	potenziometro a filo avvolto	potenciómetro m bobinado	巻線形ポテンショメータ makisengata-potenshomēta	线绕电位计 xiantao dianweiji

	English	French	German	Russian	Italian	Spanish	Japanese	Chinese
1639	work (v)	fonctionner (v)	arbeiten, bearbeiten, funktionieren, in Betrieb sein	работать, обрабатывать, действовать	lavorare, funzionare	trabajar (funcionar)	動作する dōsa-suru	工作, 运行 (动词) gongzuo, yunxing (dongci)
1640	z-transfer function	transmittance f en z	z-Übertragung-sfunktion	z-изображение	funzione di trasferimento in z	función f transformada en Z	z 変換関数 zetto-henkan-kansū	z 变换传递函数 z bianhuan chuandihanshu
1641	z-transformation	transformation f en z	z-Transformation	z-преобразование	trasformazione z	transformación f en Z	z 変換 zetto-henkan	z 变换 z bianhuan
1642	zero	zéro m	Null, Nullpunkt, Nullstellung	нуль	zero	cero m	零 zero	零,点 lingdian
1643	zero drift	dérive f du zéro	Nullpunktdrift	нулевой дрейф	deriva dello zero	deriva f de cero	ゼロドリフト zero-dorifuto	零点漂移 lingdian piaoyi
1644	zero error	taux m de dérive	Nullpunktfehler	нулевая погрешность	errore di zero	error m de cero	ゼロ誤差 zero-gosa	零点误差, 零位误差 lingdian wucha, lingwei wucha
1645	zero-order hold	bloqueur m d'ordre zéro	Halteglied nullter Ordnung	запоминающий элемент нулевого порядка	(organo di) tenuta di ordine zero	bloqueo m de orden cero, mantenimiento m de orden cero, retención f de orden cero	零次ホールド zeroji-hōrudo	零阶保持 lingjie baochi

INDEX FRANÇAIS

coupe bande	1467
filtre m à bande passante, passe-bande	99
filtre m à suppression de bande	971
filtre m actif	10
filtre m adaptatif	18
filtre m adapté	839
filtre m de Kalman	735
filtre m de Wiener	1637
filtre m harmonique, spectral	1395
filtre m linéaire	763
filtre m linguiforme	1401
filtre m non-linéaire	955
filtre m numérique	418
filtre m passe-bas	801
filtre m passe-haut	624
filtre m passe-tout	32
filtre m passif	1040
flou m	593
fonction f aléatoire	1192
fonction f aléatoire	1194
fonction f bénéfice	1045
fonction f booléenne	118
fonction f caractéristique, de critère	319
fonction f complémentaire	197
fonction f d'auto-corrélation	70
fonction f d'efficacité	1047
fonction f d'Heaviside, en échelon (de position)	1448
fonction f de commande	261
fonction f de commutation	1491
fonction f de corrélation mutuelle	325
fonction f de couplage	316
fonction f de densité de probabilité	1123
fonction f de densité spectrale	1393
fonction f de Dirac	425
fonction f de Green	612
fonction f de Lyapunov	805
fonction f de pondération	1634
fonction f de répartition de probabilité	1124
fonction f de vraisemblance	751
fonction f de Walsh	1629
fonction f échelon unitaire	1602
fonction f harmonique	617
fonction f impulsionnelle	653
fonction f propre	483
fonction f rampe, échelon m de vitesse	1186
fonction f téléologique	977
fonction f vectorielle de Lyapounov	808
fonctionnement m discontinu	103

fonctionnement m manuel	831
fonctionner (v)	583
fonctionner (v)	1639
format m, structure f d'une banque de données	355
formateur m	563
forme f canonique	128
forme f canonique de Jordan	734
forme f normale de Jordan	733
foyer m	559
frein m à courant de Foucault	480
frein m à induction	120
frein m magnétique	812
fréquence f	576
fréquence f d'échantillonnage	1298
fréquence f de coupure	326
fréquence f de coupure	330
fréquence f de coupure	599
fréquence f de coupure de phase	1059
fréquence f de résonance	1255
fréquence f de variation	301
fréquence f naturelle	1599
fréquence f naturelle amortie	333
fréquence f naturelle amortie	334
fréquence f propre	938
frottement m	582
frottement m solide	314
frottement m solide	464
frottement m solide	1434
frottement m visqueux	1626
gain m	597
gain m d'un régulateur proportionnel	1154
gain m en boucle fermée	162
gain m en boucle ouverte	795
gain m en boucle ouverte	994
gain m logarithmique	781
gain m proportionnel	1157
générateur m de fonction à diodes	424
générateur m de fonctions	586
générateur m de grandeur de référence	1222
générateur m de nombres aléatoires	1193
génération f de formes	1042
génératrice f tachymétrique	1520
génératrice f tachymétrique à courant alternatif	1
génératrice f tachymétrique, tachymètre m	1521
gestion f d'une banque de données	354
gestion f de la qualité	1176
gicleur m	731
gouvernabilité f entière	198

gouvernabilité f, commandabilité f	277
grand système m	744
grandeur f	1616
grandeur f continue	246
grandeur f d'entrée	682
grandeur f d'entrée de référence	1224
grandeur f d'entrée en forme de rampe, échelon m de vitesse	1188
grandeur f d'influence	14
grandeur f de commande	182
grandeur f de commande	825
grandeur f de référence	1226
grandeur f de rétroaction	533
grandeur f de sortie	1014
grandeur f discontinue	431
grandeur f indépendante	659
grandeur f perturbatrice	449
grandeur f réglante	308
grandeur f réglée	280
grandeur f réglée	285
grandeur f réglée de manière indirecte	663
grandeur f réglée finale	537
grandeur f réglée finale	1597
grandeur f traduite	296
graphe m de fluence	554
graphe m de fluence des signaux	1364
graphe m orienté	427
groupe m de Ward-Léonard	1630
gyromètre m	1203
gyroscope m	614
gyroscope m	615
gyroscope m d'orientation	68
gyroscope m intégrateur	706
hacheur m	157
hors-ligne, autonome	982
hyperstabilité f	638
hystérésis f	639
identifiabilité f	644
identification f	645
identification f de formes	1043
identification f des procédés	1135
image f numérique	419
impédance f	650
imprimante f	1119
imprimante f (à lignes)	758
imprimante f (à pages)	1028
imprimante f graphique	611
imprimante f laser	745
imprimante f matricielle	846

Terme	Nº
secret m des données	348
sélecteur m de signaux	1367
sélecteur m, commutateur m	1310
sensibilité f	1331
sensibilité f du système	1516
sensible	1330
séparabilité f	1333
série f convergente	292
série f de Fourier	572
série f infinie	668
servomécanisme m	1344
servomoteur m	1345
servomoteur m à membrane	398
servomoteur m hydraulique	634
seuil m	1536
seuil m de mobilité	356
seuil m de résolution	1538
seuil m logique	1537
signal m	1359
signal m analogique	46
signal m binaire	109
signal m d'action	568
signal m d'entrée	689
signal m d'entrée traduit	294
signal m d'influence	13
signal m de commande	179
signal m de commande	181
signal m de correction	293
signal m de correction	505
signal m de réaction	532
signal m de réaction	1266
signal m de référence	1223
signal m de référence	1225
signal m de sortie	1019
signal m de sortie traduit	295
signal m échantillonné	1291
signal m étalonné	1418
signal m impulsionnel	655
signal m impulsionnel	1167
signal m intermittent	719
signal m numérique	420
signal m perturbateur	448
signal m quantifié	1180
simulateur m	1373
simulation f	1371
sommateur m	1482
somme f logique (normale)	792
soupape f de sûreté	1243
souplesse f	201
sous-espace m	1479
sous-routine f	222
sous-systèmes m.pl découplés	371
sous-systèmes m.pl interconnectés	713
spécification f	1391
spectre m de fréquence	581
stabilisabilité f	1409
stabilisateur f	1412
stabilisateur m de tension	1628
stabilisation f	1410
stabiliser (v)	1411
stabilité f	1406
stabilité f à connexion partielle	1036
stabilité f absolue	2
stabilité f associative	234
stabilité f au sens de Lyapounov	807
stabilité f conditionnelle	226
stabilité f d'ambiance	496
stabilité f de configuration	229
stabilité f marginale	833
stabilité f propre	678
stabilité f structurelle	1473
stabilité f transitoire	1575
stable	1416
stationnarité f	1435
statique	1430
statisme m	462
statisme m	775
statisme m	983
stimulus m	1458
stochastique	1459
stockage m auto-organisé	1321
stockage m en mémoire adressable	241
structure f à couches multiples, feuilletée, stratifiée	917
structure f à niveaux multiples	921
structure f hiérarchisée	621
suite f markovienne	835
suite f, séquence f binaire pseudo-aléatoire	1164
supervision f	1485
supervision f d'état	1424
sur-amortissement m	1023
surcharge f	1026
surface f de commutation	1492
surveillance f	902
synchro-indicateur m	1502
synchro-machine f	1496
synchro-machine f de commande	271
synchro-machine f de puissance	1556
synchro-récepteur m de puissance	1506
synchro-récepteur m de régulation	1499
synchro-récepteur m différentiel de puissance	1504
synchro-résolveur m , synchro-trigonomètre m	1503
synchro-transformateur m de régulation	1500
synchro-transmetteur m	1508
synchro-transmetteur m de commande	1501
synchro-transmetteur m de couple	1507
synchro-transmetteur m différentiel	1505
synchro-transmetteur m différentiel de régulation	1498
synthèse f de fonctions logiques, conception f (de la logique)	785
synthèse f de signal	1368
synthèse f de systèmes	1517
système m à accès centralisés	683
système m à accès décentralisés	684
système m à déphasage minimal	885
système m à déphasage non minimal	959
système m à effet duel	468
système m à grandeurs d'entrée et de sortie multiples	916
système m à grandeurs d'entrée et de sortie multiples	934
système m à grandeurs d'entrée et de sortie uniques	1376
système m à grandeurs d'entrée et de sortie uniques	1378
système m à grandeurs d'entrée multiples	911
système m à grandeurs de sortie décentralisées	1017
système m à micro-calculateur	879
système m à niveaux multiples	922
système m à paramètres localisés	804
système m à paramètres répartis	443
système m à plusieurs degrés de liberté	832
système m à un degré de liberté	991
système m à verrouillage	1587
système m adaptatif	19
système m asservi	1343
système m auto-adaptatif	1314
système m auto-adaptatif, auto-optimisant	1320
système m auto-organisé	1322
système m bilinéaire	105
système m caché	488
système m cognitif	174
système m combinatoire	108
système m commandé	284
système m continu	245
système m d'alarme	27
système m d'apprentissage	747
système m d'échantillonnage	1302
système m d'entretien	1342
système m d'exploitation	999

Terme	Nº
système m d'interfaçage	716
système m de calcul duel	466
système m de commande automatique	78
système m de commande directe	273
système m de commande en boucle	993
système m de commande en boucle ouverte	1604
système m de commande manuelle à asservissement	828
système m de conduite	613
système m de navigation	939
système m de poursuite	1558
système m de programmation	1149
système m de recherche	1306
système m de réglage à calculateurs répartis (S.R.C.R.)	439
système m de réglage automatique	74
système m de réglage cinématique	737
système m de régulation à grandeurs d'entrée et de sortie	932
système m de régulation en boucle fermée	161
système m de régulation indirecte	662
système m de régulation par rétroaction	528
système m de régulation supervisé	901
système m de régulation supervisé manuellement	830
système m de régulation, système m asservi	272
système m de réserve, de rechange	95
système m décentralisé	363
système m déterministe	390
système m discret	432
système m discret, numérique	421
système m du deuxième ordre	1308
système m du premier ordre	544
système m échantillonné	1290
système m en boucle fermée	164
système m en temps partagé	1551
système m en temps-réel	1213
système m évolutif	1552
système m expert	515
système m flou	596
système m fondé sur l'exploitation d'une base de connaissance	738
système m hiérarchisé	622
système m homme/machine	824
système m industriel	1083
système m intelligent à base de connaissances	709
système m inverse	724
système m linéaire	767
système m markovien	837
système m modal	891
système m multidimensionnel	915
système m non-stationnaire	961

Terme	Nº
système m optimal	1005
système m permanent	1543
système m réglant	291
système m réglé par calculateur	217
système m réparti	444
système m réversible	1268
système m stochastique	1464
système m système sous-optimal	1476
systémique f	1512
table f de décision	366
table f de définition	1584
tambour m	463
taux m d'action par intégration	698
taux m d'amortissement	341
taux m de dérive	461
taux m de dérive	1644
technologie f de l'information	673
télécommande f	1244
télécommande f	1525
télécommunication f	1526
télémesure f	1527
temps m (de dosage) d'intégration	1249
temps m (de dosage) de dérivation, temps m d'activation	383
temps m caractéristique	153
temps m d'accès	5
temps m d'établissement	124
temps m d'exécution d'un système	1282
temps m de dosage d'intégration, de compensation	699
temps m de lancer	1579
temps m de montée	1270
temps m de réponse	1260
temps m de réponse	1546
temps m de retour à la normale, de recouvrement	1215
temps m de vie	750
temps m mort	359
temps m moyen avant défaillance (T.M.A.D.)	853
temps m moyen de réparation (T.M.R.)	854
temps m moyen entre défaillance (T.M.E.D.)	852
temps-réel m	1210
test m diagnostique	393
test m transitoire	479
théorème m d'ergodicité	502
théorème m de la valeur finale	541
théorème m du maximum, de Pontryagine	849
théorie f	674
théorie f catastrophe	135
théorie f de la décision	367
théorie f des automates	71

Terme	Nº
théorie f des files d'attente	1181
théorie f des perturbations	1052
théorie f des systèmes	1518
théorie f du prédicteur, de la prévision	1111
thermistance f	1531
titillation f	451
topologie f du réseau	944
train m d'impulsions	1166
train m d'impulsions	1168
train m d'ondes	1632
traitement m d'images	648
traitement m de l'information	349
traitement m de(s) données	344
traitement m du signal	1366
traitement m en parallèle	1029
traitement m en série	1080
traitement m multidimensionnel	926
trajectoire f	1559
trajectoire f d'état	1427
trajectoire f minimale	887
trajectoire f optimale	1006
transcodeur m	171
transcodeur m	1560
transducteur m	1561
transducteur m	1562
transducteur m auto-excité	90
transducteur m de mesure	861
transducteur m de mesure	862
transducteur m parallèle	1031
transducteur m série	1341
transfert m de boucle	797
transformateur m	1570
transformateur-comparateur m	409
transformateur-comparateur m	762
transformation f de Laplace	743
transformation f en z	1641
transformation f linéaire	768
transformation f modale	892
transformée f de Fourier	573
transformée f de Fourier rapide	521
transformée f de Laplace	742
transformée f inverse	727
transitoire	1571
transmetteur m	1580
transmission f des données	353
transmittance f de la chaîne d'action	506
transmittance f de la chaîne de réaction	1267
transmittance f du système	1519

DEUTSCHES VERZEICHNIS

Begriff	Nr.
Disjunktion, ODER-Verknüpfung (od. Operation)	1011
diskretes System	432
Dividierer, Teiler, Spannungsteiler	453
dominante Wurzeln	455
Doppelfeldmaschine	1403
Doppelimpuls	456
Doppelrechnersystem	466
Drahtpotentiometer	1638
Drehachse	1401
Drehmeldegeber	1509
Drehmelder, Drehfeld(system)	1497
Drehmoment, Drehimpuls	53
Drehmomentmotor	1556
Drehmomentregelung	1555
Drehmomentverstärker	130
Drehmomentverstärker	1554
Drehschwingungs- od. Torsionsschwingungsdämpfer	1558
Drehwinkelsynchro	1557
Drehzahlregler, (Fliehkraft)regler	607
Dreipunktverhalten	1533
Dreistufenlogik	1530
dreiwertige Steuergröße (plus-minus-Null)	1100
Drift, Abwanderung, Auswanderung	460
Driftrate oder Auswanderung, Auswanderungsgeschwindigkeit	461
Drossel(stelle), (Meß)blende, Durchflußbegrenzer	1264
Drossel, Luftklappe	156
Drucker	1120
Dualismus, Dualität	469
Dualitätsprinzip	872
Duhamelsches Integral	470
Duplexsteuerung	471
Durchlässigkeitsbereich	1038
Durchtrittsfrequenz (Amplitudenverlauf)	599
Düse, Austrittsdüse	973
dynamisch	476
dynamische Entkopplung	477
dynamische Programmierung	478
dynamischer Test	479
Echtzeit	1211
Echtzeitrechner	1212
Echtzeitsprache	1213
Echtzeitsystem	1214
Eckfrequenz	301
effektive Totzeit	481
effektiver Bereich	482
Egodentheorem	502
Eigenfrequenz	939
Eigenfrequenz des gedämpften Systemes	333
Eigenfunktion	483
Eigenschwingung	1324
Eigenstabilität, inhärente Stabilität	679
Eigenvektor	154
Eigenvektor	485
Eigenwert	484
Ein-Aus-Regelung, Zweipunktregelung	990
Ein-Aus-Regler, Zweipunktregler	991
Ein-Aus-Verhalten	989
Eingabeglied	686
Eingangleistungswicklung	689
Eingangs(verbindungs)matrix	590
eingangsdezentralisiertes System	685
Eingangsgerät, Eingangsbetriebsmittel	687
Eingangsgröße	683
Eingangsgröße, Eingangsvariable	691
Eingangsmatrix	688
Eingangssignal	690
Eingangssprung	1451
Eingangswicklung, Speisewicklung	1488
eingangszentralisiertes System	684
eingebettetes (auch math.) od. eingebautes System	488
Eingrößensystem	1443
Einheitsimpuls	1379
Einheitssprungfuktion	1601
Einheitssignal	1602
Einheitssprungantwort	1419
Einheitssprungantwort, Übergangsfunktion	661
Einhüllende, Hüllkurve	1603
einrichten, aufstellen, montieren	494
Einschaltglied, Einschalter	500
Einschwingzustand, transienter Zustand	970
einstellbar	1577
einstellen, abgleichen, nachstellen	23
Einstellung, Abgleichung	22
Einstellung, Einstellwert	25
Einstellvorrichtung	1354
Einstellwert, Festwert	24
Einstellzeit, Einstelldauer	1352
elektrischer Winkel, Phasenwinkel	1261
Element, Bauteil, Glied	486
Element einer Matrix, Matrizenelement	846
Emitterfolger, Impedanzwandler	487
Emulator, Kompatibilitätseinrichtung	489
endliche Ausregelzeit	490
endlicher Automat	358
endlicher Rechenautomat	543
endlicher Rechenautomat	542
Endregelglied, Stellglied	537
Endwert	540
Endwertregelung	492
Endwertregelung	1529
entkoppeltes (od. freies) Teilsystem	371
Entkopplung	372
Entkopplungsverfahren	373
entregen, abschalten, stromlos machen	375
Entscheidung	365
Entscheidungstabelle	366
Entscheidungstheorie	367
Enwertsatz	541
Erfassungsgerät, Meßfühler	387
Erfaßbarkeit, Nachweisbarkeit, Erkennbarkeit	386
Erholzeit, Ausgleichszeit	1216
erregen, einschalten	493
Erreger, Erregermaschine	511
Erregerkaskade (Hauptstromerregung)	134
Erregerwicklung, Feldwicklung	510
erreichbarer Zustand	1209
Erreichbarkeit	1208
Erwartung (statistisch)	513
Erwartungswert	514
erzwungene Schwingung	562
Expertensystem	515
exponentielle Stabilität	517
exponentielle Verzögerung	516
externer Speicher	518
Faltungsintegral	298
Farbfilter, Spektralfilter	1396
farbiges Rauschen	176
Fehler, Abweichung, Ungenauigkeit	503
Fehlererkennung	522
Fehlerkompensation	504
Fehlererkennung (… Logikfehlern)	
Fehlerort	523
Fehlersignal	505
Fehlerübertragungsfunktion	506
Fehlervorhersageverfahren, Fehlerprädiktionsverfahren	1111
Fernbedienung, Fernwirken, Fernwirktechnik (engineering)	1526
Fernsteuerung, Fernbedienung	1245
Festwertregelung (Regelkreis)	545
Flachpotentiometer	547
Flachresolver, Flach-Funktionsgeber	548
Flansch, (Drehflügel)	1614
Fliehpendelregler	143
Flipflop, bistabiler Multivibrator	549
Flügelradpumpe	1615

148

РУССКИЙ УКАЗАТЕЛЬ

INDICE ITALIANO

INDICE ESPAÑOL

Nº	Término
1347	(elemento) basculador m biestable
25	(elemento) bloqueador m
1353	(elemento) compensador m
1207	(elemento) disparador m biestable
1189	(elemento) disparador m monoestable
843	(elemento) integrador m
117	(elemento) mantenedor m
1490	(elemento) pasa-todo m
209	(elemento) programador m
16	(elemento) programador m
1312	(elemento) puerta f
255	(elemento) retenedor m
1335	(elemento) sumador m
729	(función) escalón m unitario
1217	(función) rampa
896	acceso m
30	acción correctora
1387	acción de coeficiente de trabajo
1470	acción de pulso de arranque
241	acción de tres niveles positiva-negativa
1468	acción doble
1321	acción doble con sobreposición
336	accion f
342	acción f a dos niveles
1624	acción f a dos niveles con zona muerta
329	acción f a tres niveles
1557	acción f anticipante estabilizadora
1585	acción f anticipante limitadora
337	acción f combinada
56	acción f compuesta
320	acción f continua
313	acción f correctora
340	acción f D
1600	acción f D
1023	acción f de bloqueo
1625	acción f de control
34	acción f de control limitadora
35	acción f de mantenimiento
130	acción f de muestro
158	acción f de retención
1553	acción f de derivada
36	acción f derivada
1171	acción f derivada doble
327	acción f diferenciadora
1627	acción f discontinua
131	acción f escalonada
1103	acción f escalonada múltiple
114	acción f flotante
628	acción f integradora
189	acción f integral
114	acción f integral
907	acción f intermitente
705	acción f moduladora
628	acción f multinivel
31	acción f paso a paso
1140	acción f paso a paso
1143	acción f por todo o nada
601	acción f progresiva
628	acción f progresiva
1482	acción f proporcional
1602	acción f proporcional
1186	acción f proporcional e integral
4	acción f proporcional e integral
376	acción f proporcional e integral
475	acción f proporcional e integral
1419	acción f proporcional y derivada
1099	acción f proporcional, integral y derivada
1593	acción f proporcional, integral y derivada
1594	acción flotante multivelocidad
7	acción plotante de una velocidad
1590	acción triple
1592	accionador m
1532	accionador m de mambrana
1414	accionador m hidráulico
757	accionar
203	aceleracion f
204	aceleración f angular
242	acomodación f
311	acoplamiento m
381	actuador m de solenoide
1201	actuador m final de control
627	actuador m servoasistido
254	acumulador m hidráulico
755	acumulador m intermedio
627	admitancia f
1294	adquisición f de datos
627	advance m de fase
381	adyacencia f
1201	agudeza f de resonancia
1307	ajustable
410	ajustador m
428	ajustador m del valor de consigna
1452	ajustador m del valor de consigna
929	ajustar
550	ajustar
703	ajuste m
696	ajuste m
696	alcanzabilidad
428	aleatorio
897	áigebra * matricial
918	algebra m de boole
1444	álgebra m de conmutación
1452	algoritmo * computacional
988	algoritmo m adaptativo
1150	algoritmo m autoadaptativo
1451	algoritmo m de control
1151	algoritmo m de control secuencial
1151	algoritmo m iterativo de control
1162	algoritmo m recursivo de control
1162	alimentación f modificadora
1248	alinear
1248	alisar
1159	almacen m
1161	almacenamiento direccionable por el contenido
1161	almacenamiento m
928	almacenamiento m autoorganizativo
1377	amortiguador m
1533	amortiguador m
15	amortiguamiento m de vibraciones
398	amortiguador m hidráulico
634	amortiguador m torsional de vibraciones
12	amortiguador m torsional sintonizado
3	amortiguamiento m
49	amortiguamiento m aperiódico
201	amortiguamiento m crítico
315	amortiguamiento m de coulomb
1388	amortiguamiento m relativo
538	amortiguamiento m subcrítico
1346	amortiguamiento m supercrítico
633	amortiguamiento m viscoso
123	amplidina
26	amplificación f
343	amplificador capstan
1054	amplificador chopper
20	amplificador de par
1356	amplificador m
23	amplificador m "push-pull"
24	amplificador m de corriente
1349	amplificador m de corriente
1352	amplificador m de frecuencia portadora
22	amplificador m de potencia

engranaje de rueda de aguja 1079
engranaje sumador * 1483
ensamblador * 1139
ensamblador m 59
ensamblador m 58
entrada f 682
entrada f en escalón 1450
entrada f en rampa 1188
envolvente f 494
equilibrio m 499
equilibrio m de fuerza 561
equipo m 1134
equipo m de control 259
equipo m de entrada 686
error * cuadrático medio 851
error * en régimen permanente 1441
error * transitorio 1573
error m 503
error m de cero 1644
error m de control 260
error m de cuantificación 1179
error m de histéresis 640
error m de posición 1095
error m de velocidad 1621
error m de zona muerta 357
escobillad de salida 1016
escobillas de cuadratura 1174
espacio m de estado 1426
espacio m de fases 1071
especificación f 1391
espectro m de densidad de potencia 1106
espectro m de frecuencias 581
esperanza 513
esquema m 396
esquema m de bloques 115
esquema m funcional 115
esquema m lógico 786
estabilidad f absoluta 2
estabilidad conectiva 234
estabilidad conectiva parcial 1036
estabilidad de la configuración 229
estabilidad f 1406
estabilidad f condicional 226
estabilidad f estructural 1473
estabilidad f intrínseca 678
estabilidad f marginal 833
estabilidad f propia 678

estabilidad f según Lyapunov 807
estabilidad f transitoria 1575
estabilizabilidad f 1409
estabilización f 1410
estabilizador m 1412
estabilizador m de tensión 1628
estabilizar 1411
estable 1416
estación f esclava 1385
estación f maestra 838
estación f principal 838
estacionariedad f 1435
estacionario 1437
estabilidad en elentorno 496
estado m 1420
estado m alcanzable 1208
estado m estable 1417
estado m inicial 679
estático 1430
estatismo m 462
estatismo m 775
estimación * de parámetros 1033
estimación f 507
estimación f de estado 1421
estimación f de parámetros 508
estimación f del estado 509
estimación f óptima 1004
estimador m robusto 1274
estímulo m 1458
estocástico m 1459
estrangulador 156
estructura * de bases de datos 355
estructura f arborescente 1582
estructura f jerárquica 621
estructura f multicapa 917
estructura f multinivel 921
exactitud f 6
exactitud f de control 253
excitar 493
excitatriz f 511
excitatriz f en serie 134
exclusión f 512
exponencialmente estable 517
f.a.c. 127
fabricación f asistida por computador (fac) 214
factor de transferencid diferencial 408
factor m de trabajo 474

fallo m 520
fallo m del sistema 1513
fase f 1053
fiabilidad f 1241
fiable 1242
fijar 1347
filtro * activo 10
filtro * adaptado 839
filtro * bloqueo 971
filtro * eliminador de banda 1467
filtro * espectral 1395
filtro * pasa-alto 624
filtro * pasa-bajo 801
filtro * pasa-banda 99
filtro * pasa-todo 32
filtro * pasivo 1040
filtro m adaptativo 18
filtro m de Kalman 735
filtro m de Wiener 1637
filtro m digital 418
filtro m lineal 763
filtro m no lineal 955
flip-flop m 549
foco m 559
forma f canónica 128
forma f conónica de Jordan 734
forma f normal de Jordan 733
formateador m 563
frecuencia * Fourier angular 51
frecuencia f 576
frecuencia f de corte 330
frecuencia f de cruce 326
frecuencia f de cruce de fase 1059
frecuencia f de inflexión 301
frecuencia f de la ganancia de corte 599
frecuencia f de muestreo 1298
frecuencia f de resonancia 1255
frecuencia f natural amortiguada 333
frecuencia f natural amortiguada 334
frecuencia f propia 938
frecuencia f propia 1599
freno 120
freno de corriente eddy 480
freno magnético 812
fuera de línea 982
función * Booleana 118
funcíon * criterio 319

Término	
sistema m adaptativo	222
sistema m autoajustable	371
sistema m automático de control	713
sistema m automático de control en anillo cerrado	792
sistema m autoorganizativo	1492
sistema m bilineal	1485
sistema m bimodal	1424
sistema m borroso	1012
sistema m con microprocesador	680
sistema m con un grado de libertad	604
sistema m con una sola variable de entrada y de salida	1584
sistema m con varios grados de libertad	366
sistema m continuo	1584
sistema m controlado	1521
sistema m controlado por computador	736
sistema m controlador	673
sistema m de alarma	1526
sistema m de autoaprendizaje	1525
sistema m de comunicación	1244
sistema m de control (anillo abierto)	1527
sistema m de control (anillo cerrado)	451
sistema m de control cinemático	541
sistema m de control con computador	502
sistema m de control con realimentación	674
sistema m de control en anillo abierto	1111
sistema m de control en anillo cerrado	1518
sistema m de control en anillo cerrado	367
sistema m de control multivariable	71
sistema m de desfase mínimo	1181
sistema m de desfase no mínimo	135
sistema m de doble computador	1052
sistema m de entradas centralizadas	1531
sistema m de entradas descentralizadas	393
sistema m de entradas y salidas múltiples	479
sistema m de exploración	457
sistema m de gran escala	124
sistema m de guiado	153
sistema m de lógica binaria	5
sistema m de mando en anillo abierto	383
sistema m de muestreo	699
sistema m de parámetros distribuidos	383
sistema m de parámetros localizados	1493
sistema m de parámetros repartidos	310
sistema m de primer orden	1354
sistema m de programación	1215
sistema m de salida descentralizada	1249
sistema m de seguimiento	1260

Término	
sistema m de segundo orden	19
sistema m de servicio	1314
sistema m de tiempo compartido	78
sistema m de tiempo real	74
sistema m descentralizado	1322
sistema m determinístico	105
sistema m difuso	468
sistema m discreto	596
sistema m distribuido	878
sistema m distribuido de control por computador	991
sistema m en anillo cerrado	1376
sistema m estocástico	832
sistema m hombre-máquina	245
sistema m invariante en el tiempo	284
sistema m inverso	217
sistema m jerárquico	290
sistema m lineal	27
sistema m llaves en mano	747
sistema m manual de control en anillo cerrado	183
sistema m manual de control en anillo cerrado	273
sistema m manual de control monitorizado	272
sistema m Markoviano	737
sistema m monitorizado de control	217
sistema m monovariable	528
sistema m multiacceso	993
sistema m multinivel	161
sistema m multivariable	901
sistema m no estacionario	932
sistema m operativo	885
sistema m óptimo	959
sistema m realimentado de control	466
sistema m repartido	683
sistema m repartido de control por computador	684
sistema m subóptimo	916
sistema m variable en el tiempo	1306
sistema modal	744
sistemas * basado en el conocimiento	613
sistemas * de multiprocesamiento	108
sistemas * de navegación	1604
sistemas * experto	1302
sistemas * multidimensional	443
sobrecarga f	804
sobrepaso m	443
solapamiento m	544
solapamiento m de válvula	1149
solape m	1017
sub-espacio m	1558

Término	
subprograma de computador	1308
subsistemas m desacoplados m	1342
subsistemas m interconectados m	1551
suma f lógica	1213
superficie f de conmutación	363
supervisión f	390
supervisión f de estado	596
suspensión f cardánica exterior	432
suspensión f cardánica interior	444
suspensión f de Cardán	439
tabla f de correspondencia	164
tabla f de decisión	1464
tabla f de verdad	824
tacómetro m	1543
teclado m	724
tecnología * de la información	622
telecomunicación *	767
telecontrol m	1587
telemando m	828
telemedida *	830
temblor m	830
teorema * del valor final	837
teorema m ergódico	901
teoría * de la información	1378
teoría * de predictores	911
teoría * de sistemas	922
teoría de la decisión	934
teoría f de autómatas	961
teoría f de colas	999
teoría f de las catástrofes	1005
teoría f de perturbaciones	528
termistor *	444
test m de diagnóstico	439
test m dinámico (verificación dinámica)	1476
tiempo de parada	1552
tiempo m característico	891
tiempo m característico	738
tiempo m de acceso	926
tiempo m de acción derivada	939
tiempo m de acción integral	515
tiempo m de anticipación	915
tiempo m de conmutación	1026
tiempo m de corrección	1027
tiempo m de establecimiento	1025
tiempo m de recuperación	1610
tiempo m de reposición	1024
tiempo m de respuesta	1479

tiempo m de restitución — 887
tiempo m de subida — 1168
tiempo m de subida — 1632
tiempo m de transición — 548
tiempo m de vida — 1536
tiempo m medio de reparación — 1538
tiempo m medio entre fallos — 717
tiempo m medio sin fallo — 140
tiempo m muerto — 282
tiempo m muerto — 274
tiempo m muerto — 863
tiempo m muerto — 1234
tiempo real — 589
tipo m de control — 788
tobera f — 1608
tobera f — 11
topología * de la red — 484
trabajar (funcionar) — 1494
trabajo m — 1348
transcodificador m — 1351
transductor m — 1279
transductor m de medida — 1442
transductor m magnético — 514
transductor m paralelo — 540
transductor m serie — 642
transferencia * de anillo abierto — 643
transformación f de Laplace — 694
transformación f en Z — 856
transformación f lineal — 92
transformación f modal — 1279
transformada f de Fourier — 484
transformada f de Laplace — 1539
transformada f inversa — 1609
transformada f rápida de Fourier — 275
transformador m — 1386
transformador m diferencial — 1243
transformador m diferencial — 275
transitorio m — 399
transmisión * de datos — 1081
transmisor m — 155
transmisor m de medida — 960
transmisor m paso a paso — 1243
transmitancia f — 1078
transmitancia f inversa — 1219
tratamiento m de imágenes — 1614
trayectoria * óptima — 1616
trayectoria f — 285
trayectoria f de estado

trayectoria f mínima — 1249
tren m de impulsos — 124
tren m de ondas — 1270
trigonómetro m potenciométrico plano — 1579
umbral m — 750
umbral m de resolución — 853
unidad de interfaz — 852
unidad f central de proceso — 854
unidad f controlada — 359
unidad f de control — 481
unidad f de medida — 1544
unidad f de regulación — 1210
unidad f funcional — 870
unidad f lógica — 731
valor m — 972
valor m actual — 944
valor m característico — 1639
valor m de conmutación — 472
valor m de consigna — 1560
valor m de consigna — 1561
valor m eficaz — 861
valor m en régimen permanente — 1562
valor m esperado — 1031
valor m final — 1341
valor m final ideal — 797
valor m ideal — 743
valor m instantáneo — 1641
valor m medido — 768
valor m medio — 892
valor m medio cuadrático — 573
valor m propio — 742
valor m umbral m — 727
válvula f — 521
válvula f de control — 1570
válvula f de corredera — 409
válvula f de descarga — 762
válvula f de mando — 1571
válvula f de membrana — 353
válvula f de pistón — 1580
válvula f de retención — 862
válvula f de retención — 1446
válvula f de seguridad — 1567
válvula f piloto — 725
válvula f reductora — 646
vano — 1006
variable — 1559
variable controlada — 1427

variable f activa — 14
variable f aleatoria — 1195
variable f compleja — 200
variable f continua — 246
variable f correctora — 308
variable f de conmutación — 1495
variable f de entrada — 690
variable f de estado — 1428
variable f de mando — 182
variable f de realimentación — 533
variable f de referencia — 1224
variable f de referencia — 1226
variable f de salida — 1021
variable f discontinua — 431
variable f estocástica — 1465
variable f independiente — 659
variable f indirectamente controlada — 663
variable f perturbadora — 449
variable f realimentante — 533
variable f traducida — 296
varición f paramétrica — 1034
vástago m — 1185
vector de interconexión fundamental — 591
vector m característico — 154
vector m de estado — 1429
vector m propio — 485
velocidad de auto regulación — 1205
velocidad de regulación inherente — 1204
velocidad f — 1397
velocidad f — 1620
velocidad f angular — 55
velocidad f de deriva — 461
verificacion asistida por computador — 215
verificación f automática — 86
vibrador m — 157
visualización f alfanumérica (unidad de) — 33
visualización f semigráfica — 1328
zona f de insensibilidad — 945
zona f muerta — 356
zona f muerta — 360
zona f muerta — 945

NIPPONGO NO SAKUIN

adomittansu	アドミッタンス	26
aimaisa	あいまいさ	593
akuchuēta	アクチュエータ	15
akusesu	アクセス	4
akusesu-jikan	アクセス時間	5
anarogu-dijitaru-henkanki	アナログ・ディジタル変換器	45
anarogu-keisanki	アナログ計算機	43
anarogu-seigyo	アナログ制御	44
anarogu-shingo	アナログ信号	46
anjōten	鞍状点	1284
annai-suru	案内する	1076
anpuridain	アンプリダイン	34
antei-genkai	安定限界	1408
antei-jōtai	安定状態	1417
antei-keisū	安定係数	173
antei-na	安定な	1416
antei-ryōiki	安定領域	1407
anteika	安定化	1410
anteika-fīdobakku	安定化フィードバック	1413
anteika-fīdofowādo	安定化フィードフォワード	1414
anteika-kairo	安定化回路	1415
anteika-kanōsei	安定化可能性	1409
anteika-suru	安定化する	1411
anteiki	安定器	1412
anteisei	安定性	1406
arugorizumu-gengo	アルゴリズム言語	29
asenbura	アセンブラ	58
asenbura(kigō-henkan-rūchin)	アセンブラ（記号変換ルーチン）	59
atai	値	1608
bacchi-mōdo	バッチモード	103
bacchi-seigyo	バッチ制御	102
baffa-kioku	バッファ記憶	123
baiasu-makisen	バイアス巻線	104
baio-saibanetikusu	バイオサイバネティクス	112
bakku-appu-seigyo-sōchi	バックアップ制御装置	94
bakku-appu-shisutemu	バックアップシステム	95
bakkurasshu	バックラッシュ	96
ban-ban	バンバン	101
ben	弁	1614
ben	弁	1609
ben-no-jūgōryō	弁の重合量	1610
ben-pojishona	弁ポジショナ	1613
ben-ponpu	ベーンポンプ	1615
ben-pōto	弁ポート	1612
benbetsu	弁別	434
bensen	弁栓	1611
bibun-dōsa	微分動作	410
bibun-dōsajikan	微分動作時間	383
bibun-dōsakeisū	微分動作係数	382
bibun-fido-bakku	微分フィードバック	1202
bibun-hōteishiki	微分方程式	403
bibun-kaisekiki	微分解析機	402
bibun-ki	微分器	412
bibun-yōso	微分要素	384
bibun-yōso	微分要素	411
bibundōsa	微分動作	381
binkan-na	敏感な	1330
bōdo-senzu	ボード線図	116
bōru-enbangata-sekibunki	ボールー円盤形積分器	97
bubun-kukan	部分空間	1479
bun-atsuki	分圧器	1100
bunkai-no-ikichi	分解のいき値	1538
bunkaino	分解能	1251
bunkatsu	分割	370
bunkatsu-bunken-kaiji-dendoki	分割分巻界磁電動機	1402
bunkatsu-chokken-kaiji-dendoki	分割直巻界磁電動機	1403
bunkatsu-kanshi-fidobakku	分割監視フィードバック	452
bunpu-kei	分布系	444
bunpu-moderu	分布モデル	441
bunpu-paramēta	分布パラメータ	442
bunpu-paramēta-kei	分布パラメータ系	443
bunri-kanōsei	分離可能性	1333
bunsan-gyōretsu	分散行列	1619
bunsan-kei	分散系	363
bunsan-seigyo	分散制御	440
bunsan-seigyo	分散制御	361
bunsangata-keisanki-seigyo-kei	分散型計算機制御系	439
bunsangata-seigyo-sōchi	分散型制御装置	362
bunshūki	分周器	577
burēki	ブレーキ	120
burokku-senzu	ブロック線図	115
buru-daisū	ブール代数	117
buru-kansū	ブール関数	118
busshitsu-heikō-seigyo	物質平衡制御	840

Romaji	Japanese	No.
han-i	範囲	1196
hanso-shūhasū-zōfukuki	搬送周波数増幅器	131
haretsu	波列	1632
hazumi-guruma	はずみ車	558
heikairo-dentatsu-kansū	閉回路伝達関数	165
heikairo-gein	閉回路ゲイン	162
heikairo-isōkaku	閉回路位相角	163
heikairo-seigyokei	閉回路制御系	161
heikatsuka-suru	平滑化する	1387
heikin	平均	850
heikin-chi	平均値	92
heikin-jumyō	平均寿命	854
heikin-koshō-kankaku	MTTF、平均故障間隔	852
heikin-shūri-jikan	MTTR、平均修理時間	853
heikinka-seigyo	平均化制御	93
heikō	平衡	499
heiretsu-hoshō	並列補償	1358
heiretsu-shori	並列処理	1029
heiretsu-toransudakuta	並列トランスダクタ	1031
heirūpu-kei	閉ループ系	164
heitan-kādo-potensho-mēta	平たんカードポテンショメータ	547
heitan-kādo-rezoruba	平たんカードレゾルバ	548
henatsuki	変圧器	1570
henchō	変調	898
henchō-dōsa	変調動作	897
henchōki	変調器	899
hendō-ritsu	変動率	1235
henkanki	変換器	297
hensa	偏差	391
hensa	偏差	503
hensa-shingō	偏差信号	505
henseiki	変成器	1570
hensū	変数	1616
hi-gensui-shūhasū	非減衰固有周波数	1599
hi-seigyo-buttai	被制御物体	282
hi-seigyo-sōchi	被制御装置	281
hi-senkei-na	非線形な	952
hidokishiki-tsuiji-ronri	非同期式逐次論理	63
hikaku-ki	比較器	185
hikaku-yōso	比較要素	186
hikanshō-bubun-shisutemu	非干渉部分システム	371
hikanshō-ho	非干渉法	373
hikanshōka	非干渉化	372
hiki-dashi	引き出し	1073
hinshitsu-kanri	品質管理	1176
hirei+bibun-chōsetsuki	比例+微分調節器	1158
hirei+bibun-dōsa	比例+微分動作	1159
hirei+sekibun+bibun-chōsetsuki	比例+積分+微分調節器	1160
hirei+sekibun+bibun-dōsa	比例+積分+微分動作	1161
hirei+sekibun+bibun-chōsetsuki	比例+積分+微分調整器	1163
hirei+sekibun-chōsetsuki	比例+積分調節器	1162
hirei-dōsa	比例動作	1151
hirei-dōsa-keisū	比例動作係数	1152
hirei-gein	比例ゲイン	1157
hirei-seigyo-keisū	比例制御係数	1154
hirei-tai	比例帯	1153
hireichōsetsuki	比例調節器	1155
hiritsu-seigyo	比率制御	1206
hisaishouisō-kei(-shisutemu)	非最小位相系（システム）	959
hisenkei-firuta	非線形フィルタ	955
hisenkei-hizumi	非線形歪	954
hisenkei-keikaku-hō	非線形計画法	957
hisenkei-potensho-mēta	非線形ポテンショメータ	956
hisenkei-sei	非線形性	958
hisenkei-seigyo	非線形制御	953
hishūki-gensui	非周期減衰	56
hisuterishisu	ヒステリシス	639
hisuterishisu-gosa	ヒステリシス誤差	640
hisuterishisu-mōta	ヒステリシスモータ	641
hitei	否定	940
hitei(notto)-yoso	否定（ノット）要素	970
hiteijō-kei(-shisutemu)	非定常系（システム）	961
hizumi	歪	438
hoji	保持	626
hoji	保持	778
hoji-dōsa	保持動作	627
hoji-yōso	保持要素	628
hojo-fīdobakku	補助フィードバック	1478
hokansū	補関数	197
hōkei-ha	方形波	1216
hōkei-ha	方形波	1405
hōrakusen	包絡線	494
hoshigata-senromō	星形線路網	1183
hoshō	補償	193

Romaji	日本語	No.
kakuritsu-kei(-shisutemu)	確率系（システム）	1464
kakuritsu-kinji	確率近似	1460
kakuritsu-mitsudo-kansū	確率密度関数	1123
kakuritsu-moderu	確率モデル	1121
kakuritsu-ōtomaton	確率オートマトン	1461
kakuritsu-ronri	確率論理	1120
kakuritsu-seigyo	確率制御	1462
kakuritsuteki-keikaku-hō	確率的計画法	1463
kakuritsuteki-na	確率的な	1459
kakuritu-bunpu-kansū	確率分布関数	1124
kakuteironteki	確定論的	389
kakuteironteki-shisutemu	確定論的システム	390
Kalman-firuta	カルマンフィルタ	735
kamu	カム	127
kan-shō-ki	緩衝器	329
kando	感度	1331
kanjō-senromō	環状線路網	1269
kanketsu-shingō	間欠信号	719
kanri-seigyo	管理制御	1486
kanri-shisutemu	管理システム	999
kansei	慣性	667
kansei-jōsū	慣性定数	236
kansei-masatsu	乾性摩擦	464
kansei-mōmento	慣性モーメント	900
kansetsu-seigyo-ryō	関節制御量	663
kanshi	監視	902
kanshi-fīdobakku	監視フィードバック	904
kanshi-rūpu	監視ループ	905
kanshi-yōso	監視要素	903
kanshō	干渉	315
kanshō-kansū	干渉関数	316
kanshō-kioku-sōchi	緩衝記憶装置	123
kanshō-zōfukuki	緩衝増幅器	122
kansoku-ki	観測器	981
kansōsa-gyōretsu	還送差行列	1264
kansōsa-hi	還送差比	1265
kansū-hasseiki	関数発生器	586
kansū-kinji	関数近似	584
kanwa	緩和	1238
kanzen-ka-seigyo-sei	完全可制御性	198
kasan-haguruma	加算歯車	1483
kasan-yōso	加算要素	1480

Romaji	日本語	No.
ka-kansoku-shisū	可観測指数	979
ka-kansokusei	可観測性	978
ka-kenshutsusei	可検出性	386
ka-kettei-sei	可決定性	1214
ka-seigyo-sei	可制御性	277
ka-seigyo-shisū	可制御指数	278
ka-tōtatsu-jōtai	可到達状態	1208
ka-tōtatsu-sei	可到達性	1207
kafuka	過負荷	1026
kagyaku-kei	可逆系	1268
kahen-yōryogata-ponpu	可変容量形ポンプ	1618
kai-ryōritsusei	下位両立性	458
kai-tekigōsei	下位適合性	458
kaidan-seigyo	階段制御	1445
kaidō-toransumitta	階動トランスミッタ	1446
kaidokuki	解読器	368
kaifuku-jikan	回復時間	1215
kaihō-suru	開放する	1240
kaikairo	開回路	992
kaikairo-dentatsu-kansū	開回路伝達関数	996
kaikairo-gein	開回路ゲイン	994
kaikairo-isōkaku	開回路位相角	995
kaikairo-seigyokei	開回路制御系	993
kaikimoderu-hō	回帰モデル法	875
kaikyu	階級	1200
kairo	回路	159
kairo-mō	回路網	942
kairomō-kaisekiki	回路網解析機	943
kaisō-kōzō	階層構造	621
kaisō-seigyo	階層制御	620
kaisō-seigyo	階層制御	919
kaisō-seigyo-sōchi	階層制御装置	920
kaisō-shisutemu	階層システム	622
kaitenshi	回転子	1280
kaku-shūhasū	角周波数	51
kaku-kasokudo	角加速度	49
kaku-sokudo	角速度	55
kakudo-keisoku	角度計測	52
kakudo-mōmentamu	角度モーメンタム	53
kakuichi	角位置	54
kakuritsu	確率	1122
kakuritsu-hensū	確率変数	1465

226

Page	Romaji	日本語
795	rūpu-gein	ループゲイン
796	rūpu-isōkaku	ループ位相角
794	rūpu-yōso	ループ要素
187	ryoritsusei	両立性
1178	ryoshika	量子化
1179	ryoshika-gosa	量子化誤差
1180	ryoshika-shingō	量子化信号
555	ryūryō-seigyo	流量制御
634	ryūtai-akuchuēta	流体アクチュエータ
633	ryūtai-akyumurēta	流体アキュムレータ
637	ryūtai-rirē	流体リレー
557	ryūtai-ronri-soshi	流体論理素子
635	ryūtai-zōfukuki	流体増幅器
1343	sābo-kei	サーボ系
1344	sābo-kikō	サーボ機構
1345	sābo-mōta	サーボモータ
1346	sābomōta-akuchuēta	サーボモータアクチュエータ
400	sabun-hōteishiki	差分方程式
406	sadō-haguruma	差動歯車
409	sadō-hen-atsuki	差動変圧器
583	sadō-suru	作動する
407	sadō-teko	差動てこ
401	sadō-zōfukuki	差動増幅器
331	saibanetikusu	サイバネティクス
847	saidaika	最大化
848	saidaika-suru	最大化する
1443	saidaikeishahō	最大傾斜法
1247	saigen-dekiru	再現できる
1246	saigensei	再現性
1262	saikido	再起動
1217	saikiteki-seigyo-arugorizumu	再帰的制御アルゴリズム
1108	saisa-undō-suru	蔵差運動する
1228	saisei	再生
885	saishō-isō-suii-kei	最小位相推移系
886	saishō-jikan-seigyo	最小時間制御
872	saishōjijō-nō	最小自乗法
883	saishōka	最小化
884	saishōka-suru	最小化する
540	saishū-chi	最終値
537	saishū-seigyoryō	最終制御量
541	saishūchi-teiri	最終値定理
748	saisyō-jizyō-kinji	最小自乗近似

Page	Romaji	日本語
1005	saiteki-kei(-shisutemu)	最適系（システム）
1006	saiteki-kidō	最適軌道
1003	saiteki-seigyo	最適制御
1004	saiteki-suitei	最適推定
1007	saitekika	最適化
1008	saitekika-suru	最適化にする
1531	sāmisuta	サーミスタ
1532	sanchi-dōsa	3値動作
1529	sanchi-ronri	3値論理
1533	sankō-dōsa	3項動作
1535	sankō-dōsa-chōsetsuki	3項動作調節器
1534	sankō-dōsa-seigyo	3項動作制御（PID）
28	sanpō-yōso	算法要素
1292	sanpura	サンプラ
1293	sanpuringu	サンプリング
1296	sanpuringu-chōsetsuki	サンプリング調節器
1294	sanpuringu-dōsa	サンプリング動作
1299	sanpuringu-kikan	サンプリング期間
1302	sanpuringu-shisutemu	サンプリングシステム
1298	sanpuringu-shūhasū	サンプリング周波数
1300	sanpuringu-shūki	サンプリング周期
1297	sanpuringu-yōso	サンプリング要素
1285	sanpuru	サンプル
1287	sanpuru-hōrudo	サンプルホールド
1286	sanpuru-suru	サンプルする
1288	sanpuruchi	サンプル値
1290	sanpuruchi-kei	サンプル値系
1295	sanpuruchi-seigyo	サンプル値制御
1289	sanpuruchi-seigyo	サンプル値制御
1291	sanpuruchi-singō	サンプル値信号
1099	sei-fu 3ichi-dōsa	正負3位置動作
1098	sei-no-fidobakku	正のフィードバック
6	seido	精度
120	seidō	制動
1109	seido	精度
1375	seigen-ha	正弦波
1374	seigen-yogen-potensho-mēta	正弦一余弦ポテンショメータ
30	seigō-suru	整合する
249	seigyo	制御（開回路）
248	seigyo	制御（一般用語）
247	seigyo	制御（閉回路）
255	seigyo-arugorizumu	制御アルゴリズム

234

Romanized	Japanese	No.
shūhasū	周波数	576
shūhasū-ōtō	周波数応答	579
shūhasū-ōtō-tokusei	周波数応答特性	580
shūhasū-ryō-iki	周波数領域	578
shūhasū-supekutoru	周波数スペクトル	581
shūki-ha	周期波	1049
shunkan-chi	瞬間値	694
shūsei-fīdobakku	修正フィードバック	895
shūsei-fīdofowādo	修正フィードフォワード	896
shūten-seigyo	終点制御	1528
shūten-seigyo	終点制御	492
shutsuryoku	出力	1014
shutsuryoku-burashi	出力ブラシ	1016
shutsuryoku-dentatukansū	出力伝達関数	1020
shutsuryoku-gyōretsu	出力行列	1018
shutsuryoku-hensū	出力変数	1021
shutsuryoku-jiku	出力軸	1015
shutsuryoku-makisen	出力巻線	1022
shutsuryoku-shingō	出力信号	1019
shuyou-rūpu	主要ループ	820
sibori	絞り	1263
sō-antei-toriga-yōso	双安定トリガ要素	114
sō-senkei-kei	双線形系	105
sōgo-kanshō	相互干渉	710
sōgo-sōkan	相互相関	324
sōgo-sōkan-kansū	相互相関関数	325
sōgoketsugō-gyōretsu	相互結合行列	714
sōkan-keisū	相関係数	312
sōkatsu-kanri	総括管理	1485
sokudo	速度	1620
sokudo	速度	1397
sokudo-fīdobakku	速度フィードバック	1622
sokudo-hensa	速度偏差	1621
sokutei (kenshutu)-yōso	測定（検出）要素	857
sokutei-chi	測定値	856
sokutei-fīdobakku	測定（検出）フィードバック	855
sokutei-han-i	測定範囲	859
sokutei-ten	測定点	858
sokuteiyo-hasshinki (toransu-dyūsa)	測定用発振器（トランスデューサ）	862
sonshitsu	損失	799
sorenoido-akuchueta	ソレノイドアクチュエータ	1388
sōsa-hensū	操作変数	14

Romanized	Japanese	No.
sōsa-ryō	操作量	308
sōsa-seigyo	走査制御	1304
sōsa-shingō	操作信号	13
sōsa-sōchi	操作装置	15
sōsa-suru	操作する	12
sōsa-suru	操作する	997
sōsa-yōso	操作要素	910
sōsabu	操作部	539
sōsaryō	操作量	14
sōsataku	操作卓	235
soshi-taiiki	阻止帯域	1466
sōsu-rangēji (genshi-gengo)	ソースランゲージ（原始言語）	1389
sōtaiteki-seigyo-han-i	相対的制御範囲	1237
sōtsui-hō	双対法	871
sōtsuisei	双対性	469
sūchi-seigyo	数値制御	973
sūgaku-moderu	数学モデル	841
suicchi	スイッチ	1489
suicchi-kansū	スイッチ関数	1491
suitei	推定	507
supaiku	スパイク	1399
supan	スパン	860
supan	スパン	1390
supekutoru-anaraiza	スペクトルアナライザ	1394
supekutoru-mitsudo	スペクトル密度	1392
supekutoru-mitsudo-kansū	スペクトル密度関数	1393
supin-jiku	スピン軸	1400
supurain-gata-firuta	スプライン型フィルタ	1401
suraido-ben	スライドドア	1386
suteppingu-seigyo-sōchi	ステッピング制御装置	1453
suteppu-dōsa	ステップ動作	1452
suteppu-kansū	ステップ関数	1448
suteppu-mōta	ステップモータ	1454
suteppu-nyūryoku	ステップ入力	1450
suteppu-ōtō	ステップ応答	661
suteppu-ōtō	ステップ応答	1449
suteppu-rirē	ステップリレー	1455
suteppu-suitthi	ステップスイッチ	1456
sutifunesu	ステップスイッチ	1457
syūsoku kyūsū	収束級数	292
syūsoku-seigyo	集束制御	291
ta-jiyūdo-kei	多自由度系	832

Romaji	日本語	Page
tokusei-jikan	特性時間	153
tokusei-kon	特性根	152
tokusei-kyokusen	特性曲線	149
tokusei-takōshiki	特性多項式	151
toransu-kōda	トランスコーダ	1560
toransudakuta	トランスデクタ	1562
toransudakuta-yōso	トランスデクタ要素	1563
toransudyūsa	トランスデューサ	1561
toriga-yōso	トリガ要素	1583
toruku-mōta	トルクモータ	1555
toruku-seigyo	トルク制御	1554
toruku-shinkuro	トルクシンクロ	1556
toruku-zōfukuki	トルク増幅器	1553
tōsoku-yobidashi-kioku-sōchi	等速呼び出し記憶装置	1190
tsuiji-kansōsa	逐次走査	1337
tsuiji-kirikae	逐次切り換え	1338
tsuijiteki-seigyo-arugorizumu	逐次的制御アルゴリズム	1335
tsuijū-seigyo	追従制御	560
tsuiseki-shisutemu	追跡システム	1558
tsūka-tai-iki	通過帯域	1037
tsūshin-kei	通信系	183
uzudenryū-bureki	うず電流ブレーキ	480
uzumekomi	埋め込み	656
Walsh-kansū	ウォルシュ関数	1629
Ward-Leonard-kudō	ワードレオナード駆動	1630
warikomi	割り込み	721
Wiener-firuta	ウィーナーフィルタ	1637
yakōbi-gyoretsu	ヤコービ行列	730
yamanobori-hō	山登り法	625
yoinshi-gyōretsu	余因子行列	21
yoji-kyoku	4次極	1175
yoko-jiku	横軸	1173
yoko-jiku-burashi	横軸ブラシ	1174
yōryo	容量	129
yōsekishiki-ponpu	容積式ポンプ	1097
yōso	要素	487
yosokugosa-hō	予測誤差法	1110
yosokuki-riron	予測器理論	1111
yuatsu-mōta	油圧モータ	636
yūdō-dendōki	誘導電動機	664
yūdō-kansū	誘導関数	751
yūdō-kenshutsu	誘導検出	665
yūdō-shisutemu	誘導システム	613
yudōshi	誘導子	666
yugen-ōtomaton	有限オートマトン	543
yugenjikan-seitei-ōtō(deddo-bīto-ōtō)	有限時間整定応答(デッドビート応答)	358
yukisugi-ryō	行き過ぎ量	1027
yūkō-gurafu	有向グラフ	427
yūkō-han-i	有効範囲	482
yūkō-mudajikan	有効むだ時間	481
yūkōsei	有効性	91
yūshoku-zatsuon	有色雑音	176
zatsuon	雑音	950
zen-iki-tsūka-firuta	全域通過フィルタ	32
zen-iki-tsūka-yōso	全域通過要素	31
zenkin-kinji	漸近近似	62
zero	零	1642
zero-dorifuto	ゼロドリフト	1643
zero-gosa	ゼロ誤差	1644
zeroji-hōrudo	零次ホールド	1645
zettai-antei	絶対安定	2
zetto-henkan	z変換	1641
zetto-henkan-kansū	z変換関数	1640
zōbun	増分	658
zōdai-shindō	増大振動	657
zōfuku-suru	増幅する	37
zōfuku-yōso	増幅要素	38
zōfukudo	増幅度	35
zōfukuki	増幅器	36
zukei-hyōji-sochi	図形表示装置	610
zukei-insatsu-sōchi	図形印刷装置	611

ZHŌNGWÉN SUǑYǏN

245

Pinyin	Chinese	No.
shixian, zhixing (dongci)	实现, 执行 (动词)	651
shiyan	时延	1541
shiying kongzhi	适应控制	17
shiying lüboqi	适应滤波器	18
shiying suanfa	适应算法	16
shiying xitong	适应系统	19
shiyongqi, shouming	使用期, 寿命	750
shiyouhua (dongci)	使优化 (动词)	1008
shiyu	时域	1542
shizhi	时滞	1544
shizhou	实轴	1209
shizidonghua (dongci)	使自动化 (动词)	72
shoudong bihuan kongzhi xitong	手动闭环控制系统	828
shoudong caozuo	手动操作	831
shoudong jiankong xitong	手动监控系统	830
shoudong kongzhi (kaihuan)	手动控制 (开环)	829
shoudong yingkongzhi	手动硬控制	616
shoudongde, rengong caozuode	手动的, 人工操作的	827
shoulian jishu, shoulian xilie	收敛级数, 收敛系列	292
shoulian kongzhi	收敛控制	291
shu jiegou	树结构	1582
shuaijian	衰减	65
shuaijian, shuaijianlü	衰减, 衰减率	374
shuaijian (dongci)	衰减 (动词)	64
shuaijianbi	衰减比	1477
shuaijianqi	衰减器	66
shuangchongxian maichong	双重线脉冲	456
shuanggong kongzhi	双工控制	471
shuangjisuanji kongzhi	双计算机控制	466
shuangmoshi kongzhi, shuanmotai kongzhi	双模式控制, 双模态控制	467
shuangmoshi xitong, shuangmotai xitong	双模式系统, 双模态系统	468
shuangsu kongzhiqi	双速控制器	1589
shuangwentai chufa yuanjian	双稳态触发元件	114
shuangwentai duoxie zhendangqi	双稳态多谐振荡器	113
shuangxianxing xitong	双线性系统	105
shuangzuhe kongzhi	双组合控制	465
shuchu	输出	1014
shuchu bianliang	输出变量	1021
shuchu chuandi hanshu	输出传递函数	1020
shuchu dianshua	输出电刷	1016

Pinyin	Chinese	No.
shuchu juzhen	输出矩阵	1018
shuchu raozu	输出绕组	1022
shuchu xinhao	输出信号	1019
shuchuzhou	输出轴	1015
shuju baochi	数据保持	345
shuju baomi	数据保密	348
shuju caiji	数据采集	343
shuju chuanshu	数据传输	353
shuju chuli	数据处理	349
shuju chuli	数据处理	344
shuju chuliqi	数据处理器,	350
shuju jianhua	数据简化	352
shuju jilu	数据记录	347
shuju jilu zhuangzhi	数据记录装置,	346
shuju jiluqi	数据记录器	351
shujuku guanli	数据库管理	354
shujuku jiegou	数据库结构	355
shukong	数控	973
shumo zhuanhuanqi	数模转换器	413
shunshishi	瞬时值	694
shuntai	瞬态	1576
shuntai piancha	瞬态偏差	1572
shuntai wendingxing	瞬态稳定性	1575
shuntai wucha	瞬态误差	1573
shuntai xiangying	瞬态响应	1574
shuntaide	瞬态的	1571
shunxu huicha	顺序回差	1337
shunxu kaiguan	顺序开关	1338
shunxu kongzhi	顺序控制	1334
shunxu kongzhi suanfa	顺序控制算法	1335
shuqi liju mada	竖起力矩马达	501
shuru	输入	682
shuru bianliang	输入变量	690
shuru dianyuan raozu	输入电源绕组	688
shuru huanjie, shuru yuanjian	输入环节, 输入元件	685
shuru juzhen	输入矩阵	687
shuru xinhao	输入信号	689
shuru zhuangzhi	输入装置	686
shuxue moxing	数学模型	841
shuzi jisuanji	数字计算机	415